Julian Kellner

Structural and functional studies on the human DEAD-box helicase DDX1

Julian Kellner

Structural and functional studies on the human DEAD-box helicase DDX1

Südwestdeutscher Verlag für Hochschulschriften

Impressum / Imprint
Bibliografische Information der Deutschen Nationalbibliothek: Die Deutsche Nationalbibliothek verzeichnet diese Publikation in der Deutschen Nationalbibliografie; detaillierte bibliografische Daten sind im Internet über http://dnb.d-nb.de abrufbar.
Alle in diesem Buch genannten Marken und Produktnamen unterliegen warenzeichen-, marken- oder patentrechtlichem Schutz bzw. sind Warenzeichen oder eingetragene Warenzeichen der jeweiligen Inhaber. Die Wiedergabe von Marken, Produktnamen, Gebrauchsnamen, Handelsnamen, Warenbezeichnungen u.s.w. in diesem Werk berechtigt auch ohne besondere Kennzeichnung nicht zu der Annahme, dass solche Namen im Sinne der Warenzeichen- und Markenschutzgesetzgebung als frei zu betrachten wären und daher von jedermann benutzt werden dürften.

Bibliographic information published by the Deutsche Nationalbibliothek: The Deutsche Nationalbibliothek lists this publication in the Deutsche Nationalbibliografie; detailed bibliographic data are available in the Internet at http://dnb.d-nb.de.
Any brand names and product names mentioned in this book are subject to trademark, brand or patent protection and are trademarks or registered trademarks of their respective holders. The use of brand names, product names, common names, trade names, product descriptions etc. even without a particular marking in this work is in no way to be construed to mean that such names may be regarded as unrestricted in respect of trademark and brand protection legislation and could thus be used by anyone.

Coverbild / Cover image: www.ingimage.com

Verlag / Publisher:
Südwestdeutscher Verlag für Hochschulschriften
ist ein Imprint der / is a trademark of
OmniScriptum GmbH & Co. KG
Heinrich-Böcking-Str. 6-8, 66121 Saarbrücken, Deutschland / Germany
Email: info@svh-verlag.de

Herstellung: siehe letzte Seite /
Printed at: see last page
ISBN: 978-3-8381-5119-9

Zugl. / Approved by: Heidelberg, Ruprecht-Karls-Universität, Diss., 2014

Copyright © 2015 OmniScriptum GmbH & Co. KG
Alle Rechte vorbehalten. / All rights reserved. Saarbrücken 2015

Table of contents

Summary 6

Zusammenfassung 8

1. Introduction 10

1.1 DNA- and RNA helicases 10

1.2 DEAD-box helicases 11
    1.2.1 Conserved motifs of the DEAD-box helicase core 13
    1.2.2 Conformational transitions of the DEAD-box helicase domains 15
    1.2.3 Auxiliary domains of DEAD-box helicases 17
    1.2.4 DEAD-box activities – local strand separation and ATP hydrolysis 18
    1.2.5 DEAD-box activities – more than just unwinding 24

1.3 DDX1, a unique eukaryotic DEAD-box protein 24
    1.3.1 The peculiar SPRY domain of DDX1 25
    1.3.2 DDX1 in RNA maturation processes 27
    1.3.3 DDX1's medical relevance 28

1.4 Aims of this thesis 29

2. Material and Methods 31

2.1 Materials 31
    2.1.1 Chemicals 31
    2.1.2 Buffers 32
    2.1.3 Software 34
    2.1.4 Crystallization screens 35
    2.1.5 Growth media 35
    2.1.6 Bacterial strains 36

2.2 Molecular biology 36
    2.2.1 Transformation protocol for chemically competent E. coli cells 36
    2.2.2 Amplification of DNA fragments using polymerase chain reaction (PCR) 37
    2.2.3 Site directed mutagenesis 37
    2.2.4 Preparation of plasmid DNA 38
    2.2.5 Digest by restriction enzymes and ligation of nucleotide fragments 39
    2.2.6 Agarose gel electrophoresis 39
    2.2.7 Sequencing 40
    2.2.8 Cloning of the DDX1 ORF into pET28a 40

2.2.9 Cloning of the DDX1 ORF into pET22b..41
2.2.10 Cloning of the DDX1 ORF into pGEX-4T-1...41
2.2.11 Cloning of C-terminally truncated DDX1 variants in pET28a..............41
2.2.12 Cloning of DDX1ΔSPRY in pET28a..42
2.2.13 SDM to generate a Walker A Lysine mutant of DDX1 in pET28a........42
2.2.14 Cloning of the SPRY domain in pET28a..42
2.2.15 Cloning of subunits of the HSPC117 complex into pET28a..................43

2.3 Protein biochemistry...44
   2.3.1 Recombinant protein expression..44
   2.3.2 Analysis of protein expression and purity via SDS- and native PAGE.............44
   2.3.3 Purification of full-length and truncated DDX1 variants.............................46
   2.3.4 Purification of the SPRY domain of DDX1...47
   2.3.5 Purification of HSPC117 complex components..47
   2.3.6 Co-purification and pulldown studies..48
   2.3.7 Determination of protein concentration..50
   2.3.8 Mass spectrometry...50
   2.3.9 Domain mapping by limited proteolysis experiments..................................51
   2.3.10 Electrophoretic mobility shift assay..51
   2.3.11 Helicase assay...52

2.4 Biophysical Methods..53
   2.4.1 Circular dichroism spectroscopy...53
   2.4.2 Static light scattering..54
   2.4.3 Dynamic light scattering..54
   2.4.4 Steady-state ATPase assay..54
   2.4.5 Fluorescence equilibrium titrations..55
   2.4.6 Stopped-flow measurements...57
   2.4.7 RNA affinity accessed by equilibrium titrations...57

2.5 Crystallographic methods...58
   2.5.1 Protein crystallization..58
   2.5.2 Crystal harvesting and cryoprotection...59
   2.5.3 Data collection and processing..59
   2.5.4 Phasing by molecular replacement...60
   2.5.5 Structure determination and refinement...61
   2.5.6 Homology modelling...61

3. Results 63

3.1 Generation of a stable, recombinant protein constructs...........................63

   3.1.1 Characterization of DDX1 by bioinformatic tools..63
   3.1.2 Expression and purification of recombinant DDX1.......................................66

3.1.3 Expression of the SPRY domain of DDX1 69
3.1.4 Expression of components of the HSPC117 complex and pull-down
experiments with DDX1 71

3.2 Structural studies 74

Structure of the SPRY domain 74
3.2.1 Screening for well-diffracting SPRY crystals 74
3.2.2 Phasing and refinement of the SPRY structure 77
3.2.3 Overall structure of the SPRY domain 79
3.2.4 Interface between the two β-sheet layers 81
3.2.5 Comparison of DDX1-SPRY to the structures of other SPRY domains 83
3.2.6 The SPRY domain is a conserved interaction platform 88

Structure of DDX1 93
3.2.7 Crystallization trials with full-length protein 93
3.2.8 Structural model of the entire DDX1 helicase including the SPRY domain ... 93

3.2.9 Summary of the structural studies 97

3.3 Functional characterization of DDX1 98

Biophysical properties of recombinant DDX1 98
3.3.1 Folding and oligomerization of DDX1 98
3.3.2 Characterization of conformational changes via limited proteolysis 100

Nucleotide affinity of DDX1 102
3.3.3 RNA binding observed in gel-shift assays 102
Cooperativity of RNA and ATP binding 104
3.3.4 Equilibrium titration of mant-nucleotides 104
3.3.5 Transient kinetics of mant-nucleotide binding 107
3.3.6 RNA modulates the nucleotide affinity of DDX1 110
3.3.7 Quantification of RNA affinity by spectroscopic methods 112

Enzymatic function of DDX1 116
3.3.8 ATPase activity of DDX1 116
3.3.9 Helicase activity of DDX1 118

3.3.10 Summary of the functional studies 122

4. Discussion 123

4.1. Structural studies 124
The SPRY domain structure – a modular interaction platform within DDX1 124

4.1.4 Overall structure of the DDX1 SPRY domain reveals differences to other SPRY domains 124
4.1.6 The SPRY domain might influence the enzymatic activity of DDX1 due to its central position within the helicase core 127
4.1.5 SPRY domains interact with DEAD-box helicases 129
4.1.1 A conserved SPRY surface patch as a potential protein-protein interaction region 131
4.1.2 The possible role of the SPRY domain in tethering RNA substrates to the DDX1 helicase core 133
4.1.3 N- and C-termini of the SPRY domain may constitute a flexible linker 134

4.2 Functional studies 136
Distinctive mechanistic features of DDX1 - synergistic effects of ATP and RNA binding 136

4.2.1 ATP and RNA both have the propensity to induce the "closed"-state of the helicase 136
4.2.2 Adenosine-nucleotide affinity of DDX1 – escape from an ADP stalled complex 140
4.2.3 Comparison of the cooperativity in ATP and RNA binding with other DEAD-box proteins 141
4.2.4 RNA binding - Sequence specificity 144
4.2.5 ATPase activity of DDX1 might be stimulated by external factors 145
4.2.6 Helicase deficiency of DDX1 hints towards a clamping function 147

4.3 Conclusion and Outlook 149
4.3.1 DDX1 is a novel, human DEAD-box protein involved in a plethora of cellular functions 149
4.3.2 Structural characterization of the SPRY domain 149
4.3.3 Functional characterization of DDX1 150

5. Acknowledgements 151

6. References 153

7. Appendix 167

7.1 Supplementary figures 167
7.2 Supplementary tables 184

7.3 Supplementary results on construct design, characterization and crystallization 185
    7.3.1 C-terminally truncated DDX1 constructs 185
    7.3.2 RecA-like domain 1 of DDX1 185
    7.3.3 Co-crystallization of C-terminally truncated DDX1 in complex with non-hydrolyzable ATP-analogs 186
    7.3.4 Co-crystallization of DDX1 in complex with RNA 188
    7.3.5 Transient crystal formation by using *in situ* proteolysis 190
    7.3.6 DDX1 without the SPRY insertion = DDX1ΔSPRY 191
    7.3.7 Purification of DED1 195

7.4 Comments on DDX1 crystallization experiments 196
    7.4.1 Crystallization of full-length DDX1 196
    7.4.2 Crystallization of RecA-like domain 1 of DDX1 198
    7.4.3 DDX1 conformational changes by nt binding – limited proteolysis 198
    7.4.4 Structural studies on the HSPC117 tRNA ligation complex 199
    7.4.5 The SPRY domain and the HSPC117 complex-components may control the enzymatic cycle of DDX1 200
    7.4.6 Approaches for future studies 201

7.5 General figures on helicase structure and mechanisms 204
7.6 Details on the functions of DDX1 206
7.7 List of primers 209
7.8 Numerical data analysis with Dynafit 211
7.9 List of abbreviations 214

Summary

RNA helicases are essential in all steps of RNA maturation, beginning with transcription and ending with RNA decay. They catalyze the separation of nucleic acid double strands and thereby facilitate structural remodeling. Their cellular importance is reflected in severe diseases caused by RNA helicase dysregulation. The largest group of RNA helicases is confined by the so called DEAD-box helicases of superfamily 2, characterized by the signature sequence D-E-A-D. DEAD-box helicases share a structurally conserved core of two RecA-like domains that carry signature motifs involved in ATP-binding, ATP-hydrolysis, RNA-binding and RNA-remodeling. An exceptional member of the DEAD-box protein family is the human RNA helicase DDX1 (DEAD-box helicase 1). In contrast to all other family members, DDX1 harbors a SPRY domain insertion in between the signature motifs of the helicase core. DDX1 is involved in a plethora of different RNA maturation processes and has been associated with tumor progression. Moreover due to its versatile function in RNA processing, it is hijacked by viruses for their replication. This medical relevance makes DDX1 a potential target for the development of pharmaceutics; however, such an approach would require mechanistic insights into DDX1 function.

This thesis therefore aimed at the structural and functional characterization of DDX1. The structure of the SPRY domain was determined to near atomic resolution. This structure showed two layers of concave shaped, anti-parallel β-sheets that stack onto each other. A comparison with structures of previously described SPRY domains revealed that the general fold is conserved, but the loops that mediate the interaction with partner proteins in other SPRY structures show distinct conformations and sequences in human DDX1-SPRY. A patch of positive surface charge is found in proximity of these interaction loops that is conserved within DDX1 SPRY domains and may replace the canonical protein-protein interaction surface. Based on its orientation in a homology model of DDX1, the SPRY domain possibly also enlarges the RNA binding site of the helicase core. The structural analysis is complemented by a detailed biochemical and biophysical characterization of DDX1 including the SPRY domain. Equilibrium titrations and transient kinetics with fluorescent nucleotide analogs were used to determine ATP- and ADP-binding affinities. The binding of ADP is unexpectedly tight and ADP-affinity is one of the tightest observed for DEAD-box proteins. DDX1 binds ADP tighter by a factor of almost 1000, when compared to its ATP-affinity; the latter is in the range of what has been reported for other helicases. Thus, the enzyme would be arrested in an inactive ADP-bound conformation under physiological conditions. These observations suggest that a nucleotide exchange factor – as described for the DEAD-box protein DDX19 – is necessary to resolve this ADP-stalling dilemma. Furthermore, the affinity for RNA and the influence of RNA on ATP/ADP binding was determined. Analysis of the data revealed synergistic effects of RNA and ATP binding, which were also reflected in RNA stimulated ATP hydrolysis. These observations concluded in a working model for DDX1, in which both, RNA and ATP per se have the ability to induce a conformational change to an "active" state of the helicase.

In conclusion this thesis provides the first structural information on DDX1 and by presenting the SPRY domain structure, it gives insights into the unique protein-protein interaction domain insertion. The biochemical data provide mechanistic details, show an unusual tight ADP binding and cooperativity in ATP and RNA binding, and allowed developing a model for DDX1 substrate binding.

Zusammenfassung

RNA Helikasen spielen eine entscheidende Role im gesammten Lebenszyklus eines RNA Moleküls, beginnend bei der Transkription bis hin zum schlußendlichen Abbau. Sie katalysieren das Aufschmelzen eines Nukleinsäure-Doppelstranges und helfen damit 3-dimensionale Strukturen aufzubrechen und neu zu formen. Ihre wichtige zelluläre Bedeutung zeigt sich in schweren Krankheitserscheinungen, die mit der Fehlregulation von RNA Helikasen verknüpft sind. Die sogenannten DEAD-box Helikasen der Superfamilie 2, die sich über ihre charakteristische Aminosäure-Signatur D-E-A-D definieren, stellen die größte Gruppe von RNA Helikasen dar. Sie alle teilen sich einen strukturell konservierten Kern, bestehend aus zwei Domänen, die beide dem Protein RecA ähnlich sind und die charakteristische Motife beinhalten, die in die ATP-Bindung, ATP-Hydrolyse, RNA-Bindung und RNA-Umformung involviert sind. Die menschliche RNA Helikase DDX1 (steht für DEAD-box Helikase 1) stellt ein außergewöhnliches Mitglied der DEAD-box Familie dar. Im Gegensatz zu allen anderen Familienmitgliedern wird die konservierte Grundstruktur in DDX1 durch eine SPRY Domäne unterbrochen, die sich mitten zwischen den charakteristischen Helikase Motifen befindet. DDX1 ist an einer Vielzahl unterschiedlicher RNA Reifungsprozesse beteiligt und wird auch mit der Entwicklung von Tumoren in Verbindung gebracht. Die funktionelle Vielseitigkeit in der RNA Prozessierung bringt zudem mit sich, dass Viren DDX1 für ihre Vermehrung zweckentfremden. Aufgrund der medizinischen Bedeutung, ist DDX1 ein mögliches Ziel für die Entwicklung von Pharmazeutika, allerdings würde eine solche Entwicklung grundlegende mechanistische Kenntnisse über die Funktion von DDX1 vorraussetzen.

Die Zielsetzung dieser Arbeit war daher eine stukturelle und funktionelle Charakterisierung von DDX1. Die Struktur der SPRY Domäne wurde bis zur nah-atomaren Auflösung bestimmt. Diese Struktur zeigte zwei Schichten von konkav geformten, anti-parallelen β-Faltblättern, die aufeinander gestapelt sind. Im Vergleich mit Strukturen von bereits bekannten SPRY Domänen scheint zwar die generelle Struktur erhalten zu sein, aber es fällt auf, dass die Peptid-Ketten, die in anderen SPRY Domänen den Kontakt zu Partner Proteinen herstellen, deutlich unterschiedliche Konformationen und Sequenzen in der SPRY Domäne von DDX1 zeigen. Ein Bereich mit positiver Oberflächenladung findet sich in der Nähe dieser Peptid-Ketten, der innerhalb der SPRY Domänen von verschiedenen DDX1 Homologen aus unterschiedlichen Organismen konserviert ist und eventuell die kanonische Oberfläche ersetzt, die klassischerweise für Protein-Protein Interaktion von SPRY Domänen zuständig ist. Basierend auf der Orientierung der SPRY Domäne in einem Homologie-Modell von DDX1, könnte sie eventuell auch die RNA Bindungsstelle des Helikase Grundgerüstes erweitern. Die strukturellen Ergebnisse dieser Arbeit werden von einer detaillierten biochemischen und biophysikalischen Charakterisierung der kompletten DDX1 Helikase (inklusive der SPRY Domäne) komplettiert. Mit Hilfe von Gleichgewichtstitrationen und transienten Kinetik Messungen mit fluoreszenten Nukleotid-Analoga konnten ATP- und ADP Affinitätskonstanten bestimmt werden. Die Bindung von ADP war unerwartet fest und die ADP Affinität gehört zu den festesten, die je für DEAD-box Proteine gemessen wurden. ADP bindet fast 1000fach fester als ATP, wobei die Affinität für ATP im Bereich dessen ist, was auch für andere Helikasen berichtet wurde. Die hohe Affinität zu ADP würde dafür sorgen, dass das Enzym unter physiologischen Bedingungen in einem inaktiven ADP-gebundenen Zustand stecken bleibt. Dies wiederum suggeriert, dass ein Nukleotid-Austausch-Faktor – wie bereits für das DEAD-box Protein DDX19 beschrieben – vonnöten ist um diese Blockierung durch ADP zu beseitigen. Desweiteren wurden die Affinität für RNA und der Einfluß von RNA auf die Bindung von ATP/ADP bestimmt. Die Ergebnisse zeigten synergistische Effekte zwischen der Bindung von RNA und ATP, welche auch bei der Messung der RNA stimulierten ATP Hydrolyse zur Tage treten. Auf der Basis dieser Effekte wurde ein Arbeitsmodell für DDX1 entworfen, in welchem sowohl RNA also auch ATP an sich die Eigenschaft haben eine Konformationsänderung zu einem aktiven Zustand der Helikase herbeizuführen.

Zusammenfassend lässt sich sagen, dass diese Arbeit mit der Struktur der SPRY Domäne die ersten strukturellen Informationen über DDX1 liefert und Einblicke in die einzigartige Protein-Protein Interaktions-Domänen Insertation gibt. Die biochemischen Daten liefern mechanistische Details, zeigen eine ungewöhlich starke ADP Bindung und synergistische Effekte in der Bindung von ATP und RNA, und erlauben es ein Model für die Substrat-Bindung an DDX1 zu entwickeln.

1.Introduction

1.1 DNA- and RNA helicases

Helicases are ubiquitous enzymes, initially defined by their property to use the energy obtained by nucleoside triphosphate (NTP) hydrolysis to catalyze the separation of nucleic acid double strands[1] and thus, are the central nucleic acid remodeler in the cell[2, 3]. Due to the universal need of nucleic acid structure conversion, helicases are essential in almost every process that involves DNA or RNA. A classic example for the importance of helicases in the cell is DNA replication, where they unwind long stretches of DNA in a very processive manner, which is typical for canonical helicases[4, 5].

Helicases are categorized into six superfamilies (SFs), based on biochemical activities and based on characteristic, conserved helicase sequence motifs[6, 7]. SF1 and SF2 are monomeric (or occasionally dimeric) nucleic acid dependent NTPases, whereas SF3, 5 and 6 form hexameric toroids with nucleotide binding sites at the interface between their protomers[7]. All helicase SFs share the strictly conserved P-loop motif I (Walker A) and motif II (Walker B) – essential for NTP binding and hydrolysis[8, 9], but the number and sequence of additional helicase motifs varies[7]. Helicase SF1 and SF2 that constitute the largest group of helicases harbor seven characteristic helicase motifs that show sequence similarities between both SFs[10-12] (**Figure 1.1**).

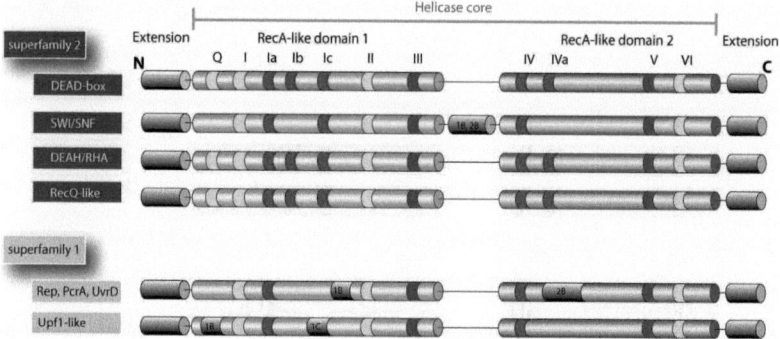

Figure 1.1 **Helicase superfamilies 1 and 2, domain organization and conserved motifs.** The basic helicase structure and position of conserved motifs is indicated for exemplary helicase families[13]. Out of the conserved motifs seven are found in both, SF1 and SF2 helicases (namely motifs I, Ia, II, III, IV, V, VI)[6]. Conserved motifs are numbered with roman letters, except for the glutamine-containing motif Q[14]. Please note that motif V of DEAD-box helicases can be further subdivided into motifs V, Va and Vb. Grey areas indicated additional domains.

SFs 1 and 2 consist of DNA- and RNA-helicases and also enzymes that can process both nucleic acids[12] (**Table 1.1**). They share a common core structure of two tandemly arranged globular α/β-domains that both resemble the fold of the RecA recombination protein (**Figure 1.2**); however, they do not show sequence similarity to RecA[15]. Despite their structural similarity, the two globular RecA-like α/β-domains that form the SF1 and 2 helicase core have very different sequences and RecA-like domain 2 lacks an ATP-binding site[16].

In contrast to SF1 of helicases that contains many processive, translocating helicases[7], SF2 is larger and functionally more diverse and has been further divided into 10 subfamilies based on sequence homology[12] (**suppl. Figure 7.5.1**). SF2 members have different nucleotide (nt) preferences and unwinding polarities (**Table 1.1**). Reported protein activities *in vitro* range from translocation coupled to unwinding, translocation without unwinding, unwinding without translocation or neither translocation nor unwinding at all[1].

Table 1.1 Enzymatic properties of individual SF1 and SF2 helicase families
n.a. means not available (adapted from Fairman-Williams, 2010)[12]

	No. of proteins			Nucleic acid preference		NTP utilization	Unwinding polarity	
	Homo sapiens	Saccharomyces cerevisiae	Escherichia coli	DNA	RNA		3' -> 5'	5' -> 3'
DEAD-box	37	26	5	-	✓	ATP	✓	✓
DEAH/RHA	15	7	2	✓	✓	ATP,GTP,CTP,T/UTP	✓	✓
NS3/NPH-II (viral)	-	-	-	✓	✓	ATP,GTP,CTP,T/UTP	✓	-
Ski2-like	7	5	2	✓	✓	ATP	✓	-
RIG-I-like	5	1	-	✓	✓	ATP	✓	-
RecQ-like	5	2	1	✓	-	ATP	✓	-
RecG-like	-	-	3	✓	-	ATP	n.a.	n.a.
SWI/SNF	28	16	1	✓	-	ATP	n.a.	n.a.
T1R	-	-	3	✓	-	ATP	✓	-
Rad3/XPD	5	2	2	✓	-	ATP	✓	-
Rep, PcrA, UvrD	2	2	4	✓	-	ATP	✓	-
Pif1-like	2	2	2	✓	-	ATP	-	✓
Upf1-like	11	5	2	✓	✓	ATP	-	✓

1.2 DEAD-box helicases

The biggest subfamily within helicase SF2 is composed of the so-called DEAD-box proteins. The name is derived from the sequence of the conserved Walker B motif (helicase motif II) Asp-Glu-Ala-Asp[17]. DEAD-box proteins show specificity for RNA substrates and exclusively hydrolyze ATP[12] (**Table 1.1**). They play key roles in RNA maturation processes[18], where a plethora of functions has been ascribed to different family members[19]. These diverse functions seem to be specific for each family member and go beyond simple RNA unwinding[20]. So far, 37 proteins of the DEAD-box subfamily have been described in *Homo sapiens*, 26 proteins in *Saccharomyces cerevisiae* and five proteins in *Escherichia coli*[12] (**Table 1.1**). The first identified DEAD-box helicase was the translation initiation factor eIF4A that consists solely of the minimal helicase sequence and serves as the prototypical DEAD-box protein[17, 21].

DEAD-box helicases share a highly conserved core (**Figure 1.2**) with an RNA- and an ATP-binding site, both sites being distributed between the two RecA-like domains[22].

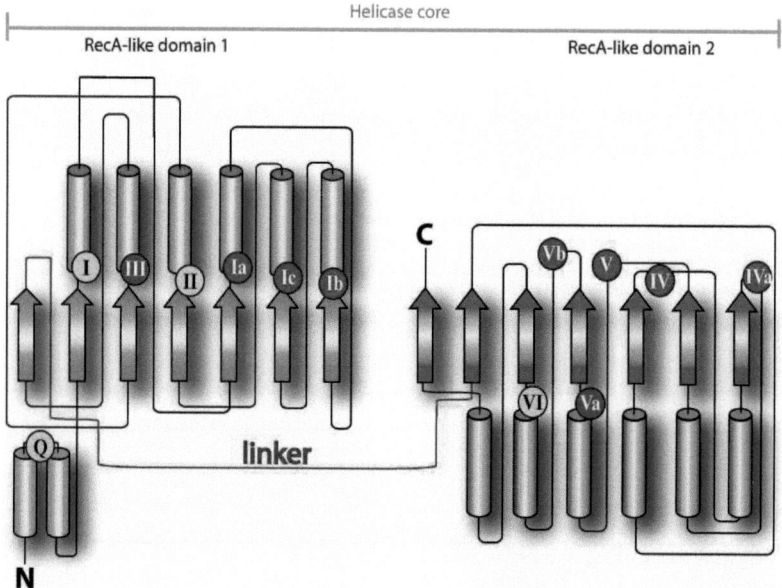

Figure 1.2 **General topology of DEAD-box proteins.** DEAD-box proteins consist of two tandemly arranged RecA-like domains like other SF1/2 helicases, but harbor distinct signature motifs. The two RecA-like domains are highly similar to each other, but only RecA-like domain 1 harbors an ATP-binding site. The conserved signature motifs at their respective positions are highlighted. The color code is according to the designated function (yellow = ATP binding and hydrolysis, red = RNA binding, purple = mediate communication between ATP- and RNA-binding site). Also see Figure 1.3 for details on the conserved motifs.

In many cases DEAD-box proteins have been found to be part of large multimeric protein complexes – well studied examples include eIF4AIII, which is part of the exon junction complex (EJC)[23, 24] and yeast Dbp5 (human DDX19), which is recruited to the nuclear pore complex (NPC)[25].

1.2.1 Conserved motifs of the DEAD-box helicase core

In addition to the seven helicase motifs shared between helicase SF1 and SF2 – motifs I, Ia, II, III, IV, V, VI[6] (see **Figure 1.1**) – the two globular RecA-like domains of DEAD-box proteins (**Figure 1.2**) harbor further family specific signature motifs. These include motifs Q, Ib, Ic, IVa, Va, V resulting in an overall number of at least 13 conserved sequences[19] (**Figure 1.3 a**). Seven of the conserved motifs are located on the N-terminal-, six motifs on the C-terminal RecA-like domain.

The motifs can be separated into three groups, according to their function in ATP binding/hydrolysis (depicted in yellow in **Figure 1.3**), RNA binding (depicted in red in **Figure 1.3**) or in mediating communication between the ATP and RNA binding site (depicted in purple in **Figure 1.3**) [18].

ATP binding and hydrolysis is mediated by motifs Q, I and II that belong to RecA-like domain 1 and motif VI that belongs to RecA-like domain 2[26, 27]. The Q-motif contains a conserved glutamine that coordinates the adenine base moiety and thereby confers ATP specificity[14] (**Figure 1.3**). Typically a conserved phenylalanine of the Q-motif additionally stacks with the adenine base moiety[28]. Next in the primary structure is the highly conserved P-loop motif I=Walker A motif with the sequence G-S/T-G-K-T, which is DEAD-box family specific[17] and in contrast to the classical P-loop motif G/A-X-X-X-X-G-K-S/T[10]. The Walker A motif interacts with the phosphate groups of ATP/ADP and coordinates the Mg^{2+} ion in the ATP binding site[29]. More specifically, the ε-amino group of the Walker A lysine interacts with the β-phosphate group and the hydroxyl-group of the serine coordinates the Mg^{2+} ion octahedrally[15] (**Figure 1.3 b and c**). Helicase motif II=Walker B motif corresponds to the conserved D-E-A-D sequence and is essential for ATP hydrolysis[30]. The first aspartate residue coordinates the Mg^{2+} ion with its carboxyl-group and the glutamate residue acts as the catalytic base and activates a water molecule for nucleophilic attack on the γ-phosphate group[19] (**Figure 1.3 c**). Finally, motif VI of RecA-like domain 2 provides two arginine fingers that coordinate the phosphate groups of bound ATP and it also interacts with the nucleobase[31]. It thereby complements the ATPase site that otherwise only consists of residues from RecA-like domain 1 (**Figure 1.3 b**).

RNA binding is mediated by motifs Ia, Ib and Ic that cluster at the surface of RecA-like domain 1 and constitute the first part of an RNA binding site[31] (**Figure 1.3 b**). RecA-like domain 2 harbors conserved motifs IV, IVa, V and Vb, which constitute the second part of the RNA binding site (**Figure 1.3 b**).

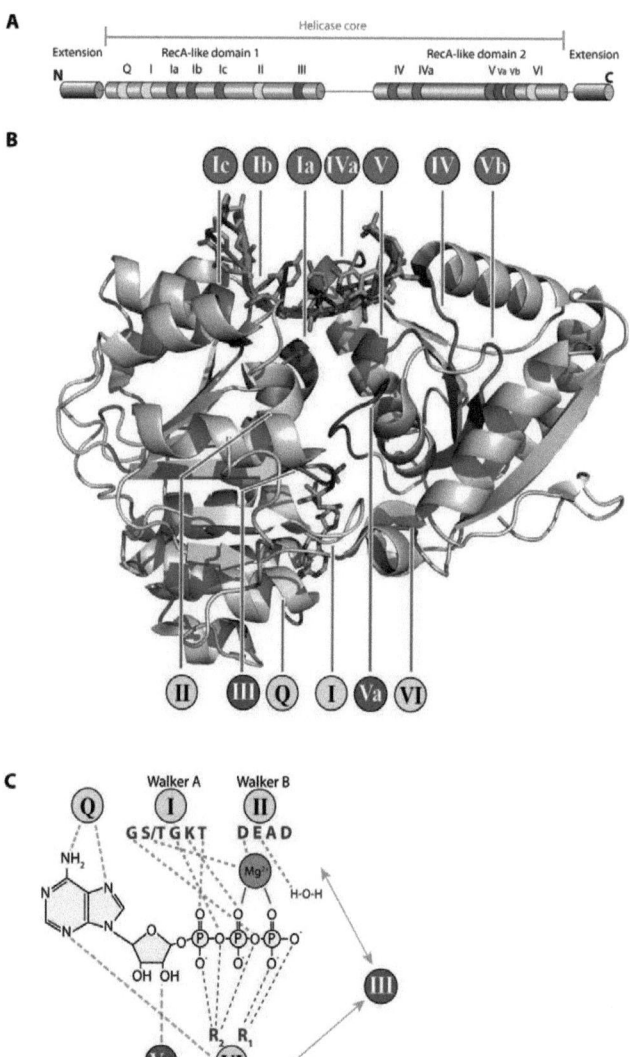

Figure 1.3 **DEAD-box helicase core and conserved motifs. A**, A schematic representation of the helicase core with the location of the conserved motifs is shown. Helicase motifs are colored according to their function (yellow = ATP binding and hydrolysis, red = RNA binding, purple = mediating communication between ATP- and RNA-binding site). **B**, Ribbon representation of the structure of the helicase core of eIF4AIII[24] (PDB entry 2J0S) is shown with the conserved motifs colored according to A, ATP and RNA represented as stick-models. **C**, A schematic representation of the interactions of ATP with the conserved motifs in the ATP binding site is shown[15, 19].

Introduction

Communication between the ATP and RNA binding site is mediated by motifs III and Va[32]. Motif III of RecA-like domain 1 forms an extended network of hydrogen-bond interactions and is essential for RNA unwinding activity[30]. It senses bound ATP and signals to the RNA binding site[13]. In RecA-like domain 2, motif Va recognizes the ribose moiety of bound nucleotides and is also involved in coordination between ATP hydrolysis and RNA binding[12].

1.2.2 Conformational transitions of the DEAD-box helicase domains

For a long time only apo or ADP bound structures of DEAD-box proteins have been available[13, 32]. These structures show that the two RecA-like domains of the DEAD-box helicase core are connected by a short flexible linker[33]. The relative orientation of the two RecA-like helicase domains varies significantly between different DEAD-box protein apo structures, indicating large domain flexibility[34]. ADP binding does not seem to have an influence on domain orientation[26, 29].

Back in 2006 a breakthrough in the field came with the first structure of a DEAD-box helicase bound to a single-stranded RNA (ssRNA) and a non-hydrolyzable ATP analog[31]. Since then several similar substrate-bound complexes of different DEAD-box proteins have been reported[23-25, 28, 35]. Interestingly, all ssRNA-NTP-helicase complexes show a virtually identical orientation of the two RecA-like domains with respect to each other, the so-called "closed"-state. In contrast, an "open"-state is represented by the apo form, where the relative domain orientations are not exactly defined[34]. Thus, two distinct states of DEAD-box helicases are structurally characterized[13]. In the absence of NTP and/or RNA substrate DEAD-box proteins are in an "open"-state with a flexible arrangement of the two RecA-like helicase domains, whereas they adopt a rigid "closed"-state upon binding of non-hydrolyzable ATP-analogues and RNA.

Structures of the ternary DEAD-box protein, RNA, ATP complex have also helped to assign the functional contribution to individual helicase motifs in substrate binding[31]. In the "closed"-state, the two helicase domains come in close proximity and form a cleft, which establishes the functional ATPase site between the domains[31]. RNA molecules bind across the surface of both RecA-like domains opposite of the ATP-binding cleft (**Figure 1.3**). They are bound by the conserved helicase motifs that contact five nucleotides of the RNA[34]. Interactions exclusively rely on the RNA sugar phosphate backbone, thereby confering the low sequence specificity of DEAD-box helicases[34]. Despite this sequence promiscuity, DEAD-box protein - oligonucleotide interactions are highly RNA specific, which is mediated by interactions with the 2'-hydroxy-group of the sugar moiety[24].

The most important observation from structures of DEAD-box helicase in the "closed"-state was a characteristic bend of the bound single-stranded RNA (ssRNA)[31] (**Figure 1.4**). This distorted conformation of the RNA is caused by steric constraints imposed by RecA-like domain 1[32]. The observed RNA conformation would interfere with the Watson-Crick base pairing of double-stranded RNA (dsRNA), suggesting that tethering and subsequent bending of RNA by DEAD-box proteins leads to melting of the double-stranded region and finally to strand separation[31].

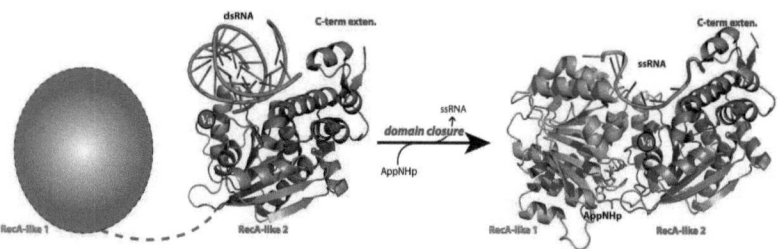

Figure 1.4 **DEAD-box protein dsRNA binding and unwinding by bending.** On the left panel, a ribbon representation of the structure of the isolated RecA-like domain 2 (including a C-terminal extension domain) of the yeast DEAD box helicase Mss116p bound to dsRNA is shown[36]. RecA-like domain 1 is not resolved in this structure and indicated by a green blob. On the right panel, a ribbon representation of the structure of the entire helicase core (RecA-like domains 1, 2 and the C-terminal extension) of Mss116p bound to ssRNA is shown[35]. It has been suggested, that upon formation of the "closed"-state of the helicase both RecA-like domains come together and the second RNA strand is displaced, i.e. the dsRNA is melted, induced by sterical constraints via strand-bending[32]. Motif Va interacts with double-stranded substrate in the "open"-state, but probably alters its position in the "closed"-state to help form the ATP-binding site[36]. PDB codes are as follows: Mss116p RecA-like domain 2 and C-terminal extension with dsRNA (4DB2[36]); Mss116p helicase core and C-terminal extension with ssRNA/AppNHp (3I5X[35]).

The hypothesis of an "unwinding by bending" mechanism of DEAD-box proteins was further substantiated by a recent study that showed that dsRNA and ATP bind independently to RecA-like domain 2 (**Figure 1.4, left side**) and RecA-like domain 1, respectively, in the "open"-state[36]. As suggested before[32], upon closure of RecA-like domains 1 and 2 (formation of the "closed"-state), parts of the bound strand are bent, i.e. the A-form conformation is destabilized, which leads to displacement of the second strand by the steric constraints that disturb the Watson-Crick base pairing[36] (**Figure 1.4, right side**). Interestingly, ssRNA forms additional contacts with RecA-like domain 1 after strand melting, which provides further energy for domain closure and the unwinding reaction[37]. Comparison of a structure of the isolated RecA-like domain 2 bound to dsRNA and the structure of complete helicase core (RecA-like domains 1 and 2) bound to ssRNA also shows significant structural changes in motif Va (**Figure 1.4**). It has been suggested that motif Va initially contributes to duplex RNA binding and rearranges to help form the ATPase active site in the "closed"-state[36].

So far, no structure of a binary complex of a DEAD-box protein bound to ATP or bound to RNA has been determined. For the SF1 helicase Upf1 though (which is not a DEAD-box protein), structures of a binary protein-nucleotide complex show a conformational change and a domain closure in the AppNHp bound state when compared to the ADP bound state (both in the absence of RNA)[38].

Introduction

1.2.3 Auxiliary domains of DEAD-box helicases

In addition to the conserved helicase core many DEAD-box proteins harbor auxiliary domains that contribute to their specific functions. In contrast to other helicase-families these domains are generally not inserted in the two RecA-like domains, but form N- and C-terminal extensions[12]. Those extensions function as protein-protein or protein-RNA interaction platforms[19]. A protein-RNA interaction platform has for instance been observed in the structure of the C-terminal extension domain of *Bacillus subtilis* YxinN, which shows the fold of an RNA recognition motif (RRM)[39]. The C-terminal extension of *Thermus thermophilus* HerA protein contains both a protein-protein dimerization domain and an RRM[40-42] (**Figure 1.5**). Another example is the C-terminal domain of *S. cerevisiae* Mss116p that binds to the RNA, in addition to the helicase core and introduces a second bend in the RNA; thereby it contributes to the unwinding activity of the core[35] (**Figure 1.4**, right side).

Protein-protein interactions have for instance been described for *Drosophila melanogaster* VASA, whose N-terminal extension interacts with GUSTAVUS (GUS) while its C-terminal extension interacts with OSKAR protein[43] and eIF5B[44].

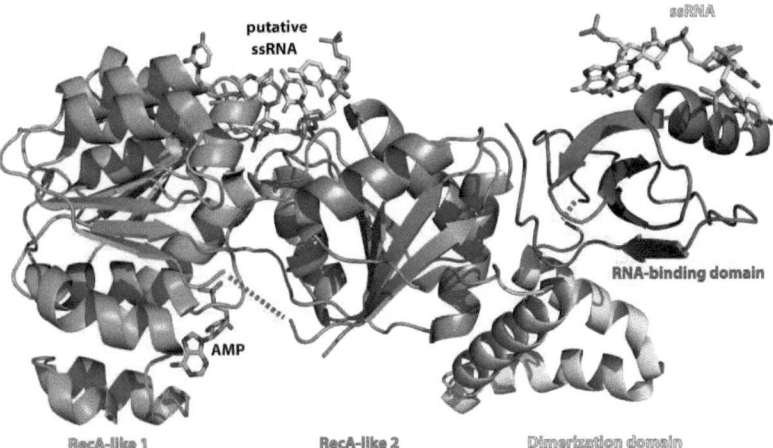

Figure 1.5 **C-terminal extension domains of *T. thermophilus* HerA adopted from Klostermeier et al.**[40]. Similar as described in Rudolph and Klostermeier (2009)[41] and in Klostermeier and Rudolph (2009)[40], a model of the entire HerA helicase was assembled, based on individual structures. The structures of the *T. thermophilus* HerA protein N-terminal domain[45] (PDB entry 2GXS) and of the C-terminal domain with the dimerization domain and the RNA binding domain=RRM[41] (PDB entry 3I32) were superposed on the structure of the closed state of *D. melanogaster* VASA in complex with AppNHp and ssRNA[31] (PDB entry 2DB3). Then the structure of the RRM in complex with ssRNA[42] (PDB entry 4I67) was superposed on the C-terminal domain. This model of the entire HerA helicase was constructed by using "secondary structure matching" in COOT[46]. The N-terminal domain of HerA was crystallized in complex with AMP without RNA[45], and the position of ssRNA is hypothetical and inferred from the VASA structure. Linkage of the separated crystal structures in the model of the entire protein is indicated by dashed purple lines. The C-terminal domain of HerA contains both a dimerization domain and an RRM. The RRM can tether RNA substrates or can bind large RNA substrates that span both RecA-like domains and the RRM[42, 47].

Introduction

In addition to binding to substrates or interaction partners, auxiliary domains of DEAD-box helicases do also have regulatory functions, e.g. in *H. sapiens* DDX19, the N-terminus controls ATP hydrolysis by conditionally displacing a catalytical arginine finger[28]. Interestingly this N-terminus is locked in an non-inhibiting conformation in the structure of DDX19 complexed with its interaction partner Nup214, reflecting modulation of DEAD-box activity by external factors[48]. Individual orientations of the auxiliary domains that have been determined so far, were all different with respect to the helicase core[19].

1.2.4 DEAD-box activities – local strand separation and ATP hydrolysis

DEAD-box proteins use a different mechanism of substrate unwinding, when compared to all other RNA helicases[49] (**Figure 1.6 a**). They are unprocessive and do not show strict unwinding polarity[11]. Whereas classical helicases unwind RNA-duplexes by translocation on one of the strands, DEAD-box proteins employ an unwinding mode termed "local strand separation"[50-52] (**Figure 1.6 b** and **Figure 1.8**).

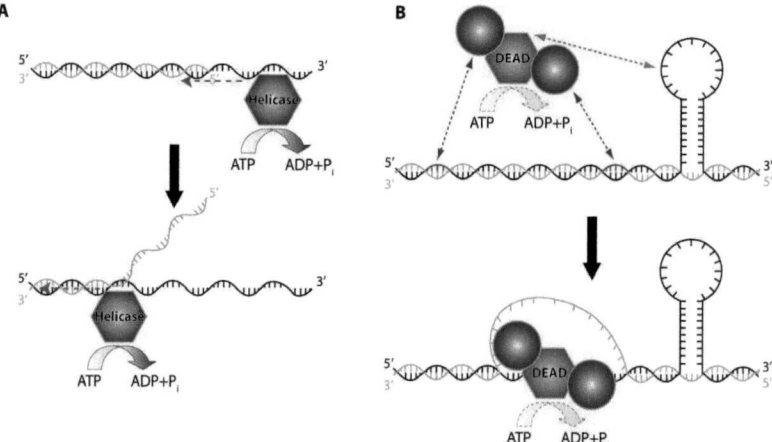

Figure 1.6 **Unwinding mechanism of translocating helicases differs from DEAD-box proteins. A**, Classical translocating helicases unwind long stretches of duplex substrate with high processivity. They are tethered to their substrate by single-stranded overhangs and translocate with a defined polarity, thereby displacing the non-guidance strand. **B**, DEAD-box proteins can be tethered to their substrate by interacting factors (depicted in purple) or by certain structural elements. They unwind short duplexes in a non-processive manner and don't necessarily require the energy of ATP hydrolysis. In contrast to classical helicases, they initiate the unwinding internally in the dsRNA[20].

DEAD-box proteins do not need a single-stranded overhang, but load on the double-stranded region and then open the strands just locally[52] – therefore the term "local strand separation" (**Figure 1.6 b**). Loading on the double-stranded region can be facilitated by ssRNAs that do not even have to be in spatial proximity to the dsRNA substrate[51]. Many family members are only capable of unwinding

Introduction

duplex RNA with 14 or less base pairs, corresponding to less than two helical turns[19]. Increasing stability of the duplex substrate (via GC content) is proportional with decreasing unwinding efficiency[53-55]. The non-processive unwinding mechanism of DEAD-box helicases is well adapted for local remodeling processes in RNA and ribonucleoprotein (RNP) processing[13]. Unwinding of short double-strands prevents large scale dismantelling of complex RNA or RNP structures and is well suited for RNA duplexes, commonly formed in the cell that rarely exceed one helical turn[19]. The mechanism of "local strand separation" can be explained by the "unwinding by bending" hypothesis with the conformational changes between the "open"- and "closed"-state of the helicase[32]. The few base pairs of duplex that are separated by bending of one RNA strand and exclusion of the other strand by domain closure are sufficient to locally open the double strand, which leads to strand separation and unwinding only in short duplexes[36].

Cycling between the "open"- and the "closed"-state is inseparably coupled to ATP hydrolysis[17]. DEAD-box proteins hydrolyze ATP, though with catalytic rates that vary between 0.5 min^{-1} to 300 min^{-1} [56-58]. For many family members the intrinsic low ATPase-rate is stimulated significantly by ssRNA[59, 60]. The degree of stimulation of ATP hydrolysis by ssRNA ranges from less then ten-fold to several orders of magnitude for different proteins[37]. However, stimulation of ATP hydrolysis by ssRNA does not necessarily mean that a given protein is able to unwind duplex nucleic acids, since not for all proteins that showed an RNA stimulation an unwinding activity could be shown[12].

The ATPase cycle can be regulated by interactions with co-factors. For instance, the human eIF4A is locally activated in its ATPase activity by the scaffolding protein eIF4G that induces a half-open conformation of eIF4A and thereby accelerates phosphate release after ATP hydrolysis[55, 61, 62] (**Figure 1.7 a** and **b**). In contrast, the natural tumor suppressor protein PDCD4 binds two eIF4A protomers and traps them in an inactive conformation, thereby inhibiting the enzymatic activity[63, 64] (**Figure 1.7 c**).

Introduction

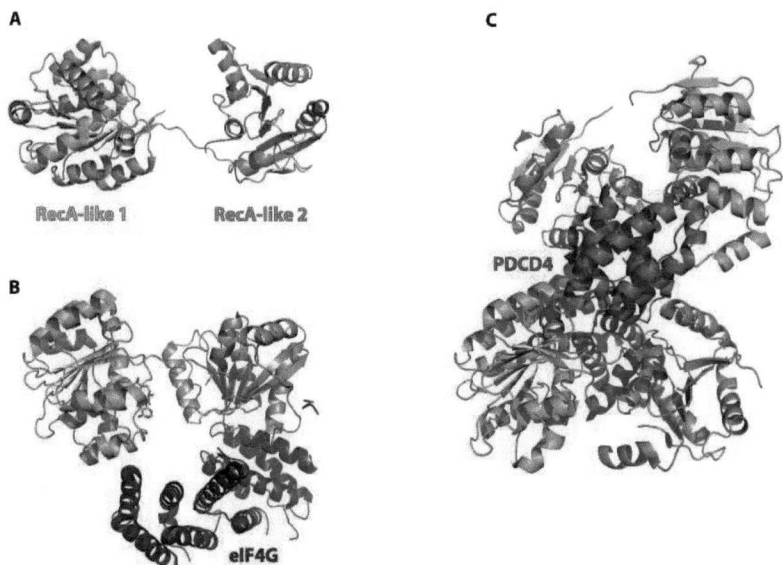

Figure 1.7 **Regulation of the ATPase cycle of eIF4A by co-factors.** Human eukaryotic initiation factor 4A (eIF4A) is modulated in its ATPase cycle by interacting factors. **A**, apo-structure of eIF4A[65] is depicted with the color code as in previous figures. **B**, The large scaffolding factor eIF4G binds both RecA-like domains and induces a half-open form[61] that promotes phosphate and RNA release[62]. **C**, The natural tumor suppressor protein "programmed cell death 4" (PDCD4) inhibits translation by binding to eIF4A[66]. Two protomers of eIF4A bind to one molecule PDCD4, which traps the DEAD-box protein in an inactive conformation and prevents its binding to eIF4G[63, 64]. PDB codes are as follows: (A) – apo eIF4A (1FUU), (B) – eIF4A/eIF4G-complex (2VSO), (C) – eIF4A/PDCD4-complex (2ZU6).

Similar to eIF4A, ATP turnover by yeast Dbp5 (= human DDX19) is stimulated by the export factor Gle1[25]. Furthermore, a nucleotide exchange factor (NEF) plays a role in the Dbp5 regulation as the nucleoporin Nup159 (= human Nup214) acts as an ADP release factor (**Figure 4.3.1**) – this shows the presence of NEFs in RNA helicase regulation[67]. In contrast to the activation observed for Dbp5, the protein eIF4AIII in the exon junction complex is inhibited by the MAGOH-Y14 complex, which prevents release of the inorganic phosphate after hydrolysis and thereby keeps the helicase bound to the RNA[68].

Whereas conventional helicases processively translocate through their substrate and consume several ATP molecules for the complete melting[69, 70], DEAD-box proteins require only a single ATP molecule for stand separation[71]. However, their catalytic cycle is not necessarily always productive i.e. can lead to futile ATP hydrolysis without dsRNA strand separation[72]. Interestingly, ATP binding is required for strand separation (**Figure 1.8**– helicase cycle), whereas hydrolysis can be dispensable[73, 74]. Hydrolysis of

ATP is finally required to release the RNA from the DEAD-box protein, to complete substrate turnover and to reset the helicase for another unwinding cycle[57]. In general, ATP hydrolysis is fast and reversible, whereas the release of inorganic phosphate or ADP is the rate limiting step for the catalytic cycle[19, 57].

So far, detailed kinetic analysis of DEAD-box proteins is limited to *E. coli* DbpA and *S. cerevisiae* Mss116p and these studies were restricted to constructs of the helicase core, not containing large extensions or insertions[57, 58, 72]. Additional studies have characterized the ATPase cycle of *Mus musculus* eIF4A, *E. coli* Ded1p, *Bacillus subtilis* YxiN (a DbpA homologue) and *Thermus thermophilus* HerA[59, 75-77]. These studies show that ADP is often bound with higher affinity than ATP; sometimes up to one order of magnitude more tightly[34, 60]. In some situations cooperative effects of RNA and ATP binding have been observed[75], however, this is not generally true for all DEAD-box helicases. For instance in the case of DbpA, cooperativity is controversely discussed as it is in contrast to kinetic data[78-80].

A combination of the observations on local strand separation and ATP hydrolysis finally concluded in a general model for the overall unwinding cycle of DEAD-box proteins[37] (**Figure 1.8**). The cycle starts with initial binding of RecA-like domain 2 to the ds substrate[36]. ATP binding to RecA-like domain 1 induces the conformational transition to the "closed"-state in which dsRNA is separated locally and a functional ATPase site is formed[31] (**Figure 1.8, step 2**). Local melting of a few base pairs of the duplex can lead to opening of longer regions and finally to complete strand separation, resulting in the "closed"-state with bound ssRNA and ATP as observed in many crystal structures[23, 24] (**Figure 1.8, step 3**). Importantly, after the unwinding, ATP is hydrolyzed (**Figure 1.8, step 4**), leading to a preference for the "open"-state in which P_i and ADP as well as RNA are released[60]. At this point, the helicase is reset for another cycle of duplex unwinding[1] (**Figure 1.8, step 5**). Importantly in the cellular environment DEAD-box associated factors can bind the ssRNA and thus, prevent an immediate reannealing[19].

Introduction

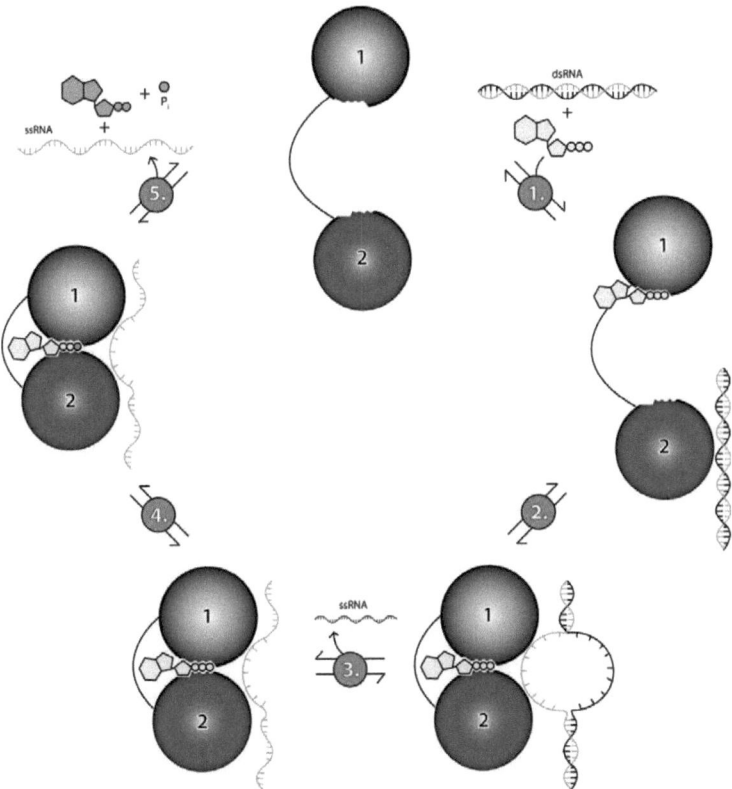

Figure 1.8 **Unwinding cycle of DEAD-box proteins.** The cycle of unwinding by 'local strand separation' by DEAD-box proteins is depicted schematically. RecA-like domain 1 binds ATP and RecA-like domain 2 binds dsRNA substrate (step 1), which leads to domain closure. Both domains bind their respective substrate independently of each other[36]. Domain closure leads to melting of the Watson-Crick base pairing in the bound region due to sterical constraints and strand bending[31] (step 2). In a productive helicase cycle, the displaced RNA strand dissociates (step 3). ATP hydrolysis occurs after the unwinding event to reset the enzyme (step 4). Dissociation of the ATP hydrolysis products and the RNA finally resets the enzyme.

1.2.5 DEAD-box activities – more than just unwinding

Besides the basic helicase function of separating nucleic acid double strands, DEAD-box proteins can exert a range of other biochemical activities. They can displace proteins from RNA in an ATP-dependent way, independent of duplex unwinding[81]. However, only proteins that interact with less than eight nucleotides have been reported to be displaced, most likely due to the non-processive nature of DEAD-box proteins[82, 83].

Another activity of DEAD-box proteins is RNA strand annealing, which, at first glance, seems to contradict the duplex unwinding activity (**suppl. Figure 7.5.2**). However, annealing has only been shown *in vitro*[84] and is not the reverse of the unwinding reaction since it is ATP independent[84-86]. For DDX21 ATP-independent RNA folding has been reported, which may also work via an annealing mechanism[87] (**suppl. Figure 7.5.2 a**). The annealing activity combined with unwinding activity allows for large scale conversion of RNA structures[88]. This is important for the chaperone-like function of DEAD-box proteins on certain RNA molecules[20, 89] (**suppl. Figure 7.5.2 b**).

Furthermore, DEAD-box proteins can remain bound to RNA for prolonged periods of time, a function referred to as RNA clamping[68]. Such clamping has been reported for eIF4AIII, which functions as an assembly platform for the exon junction complex (EJC)[23, 24, 68]. Similarly a clamping function has recently been reported for the yeast protein Mss116p that forms nucleotide-dependent, long-lived RNA-bound complexes[37, 90].

1.3 DDX1, a unique eukaryotic DEAD-box protein

The human RNA helicase DDX1 – for DEAD-box protein 1 (originally called HuDBP-RB = human DEAD-box protein amplified in Retinoblastoma cells) – is a 80 kDa DEAD-box protein localized both in the nucleus and in the cytoplasm, depending on cell-type[91]. What differentiates DDX1 from all other DEAD-box helicases is the presence of a long SPRY-like domain insertion directly within the helicase core[92]. Besides for its unique domain composition, DDX1 is also interesting for its manifold involvement in RNA processing pathways in the cell (see section 1.3.3) and is of high medical relevance due to its role in tumor progression and HIV-1 replication[93, 94]. DDX1 is an essential protein, as knockout in *D. melanogaster* results in a lethal phenotype[95].

Introduction

1.3.1 The peculiar SPRY domain of DDX1

DDX1 contains a long (constitutes almost 30% of the residues of the protein) insertion within RecA-like domain 1, which is located between conserved helicase motifs I and Ia in the polypeptide chain[92] (Figure 1.9). In contrast to an average 20-40 residues in other DEAD-box proteins, the insertion seperates the two motifs by around 240 residues in the primary sequence[96].

Figure 1.9 **The prominent SPRY insertion in DDX1.** DDX1 is depicted schematically with the two canonical RecA-like domains, depicted in green (domain 1) and blue (domain 2) respectively. Additional segmentation at the ends marks N- and C-terminal extensions that are not part of the conserved helicase core. The SPRY domain is inserted prominently between conserved motifs I (=Walker A) and Ia. Additionally the position of motif II (=Walker B) is marked to show the spatial proximity to the SPRY domain.

Sequence alignment indicated that the insertion corresponds to a so-called SPRY domain[92]. SPRY domains are modular insertion domains, which were originally identified as a sequence repeat in the dual-specificity kinase splA (=SP) and Ca^{2+}-release channel ryanodine (=RY) receptors[97]. SPRY domains are highly conserved in their amino acid sequence[98]. A significant number of SPRY domain containing proteins harbor an N-terminal domain fusion with a protein domain termed PRY[99] (named after its association with SPRY domains). The fused PRYSPRY domains have a strong structural homology to the so called B30.2 domains, however, on the sequence level only the SPRY domains are conserved and the PRY domains are different[100]. It has therefore been proposed that PRYSPRY and B30.2 domains have derived from the same ancestral gene, but the first part of the gene has diversified greatly beyond the level of sequence similarity[100, 101]. This led to the recent renaming of PRYSPRY and B30.2 domains to the collective labeling as SPRY domains in general[98, 101].

SPRY domains are present in more than 150 human and ~80 murine proteins of different protein families that cover a wide range of functions[99]. They constitute protein-protein interaction modules, as could be shown by interaction studies[102-104]. Structural studies of SPRY domains of the mammalian SOCS box family of E3 ubiquitin ligases (SPSB) in complex with interacting regions of complex partners helped to identify the so-called surface A interaction site[101, 105]. Surface A is established by five extended loops[105, 106]. Initially, comparison of the structure of the orthologous drosophila protein GUSTAVUS in apo and in a peptide bound complex revealed that the five extended loops of surface A create a pre-formed pocket [100, 101]. This was soon corroborated by structures of human SPSB proteins in apo or in different peptide bound complex forms[105] (**Figure 1.10**). Notably GUSTAVUS interacts tightly with the DEAD-box helicase VASA and the interacting peptide, used for structural studies was derived from this RNA helicase[44].

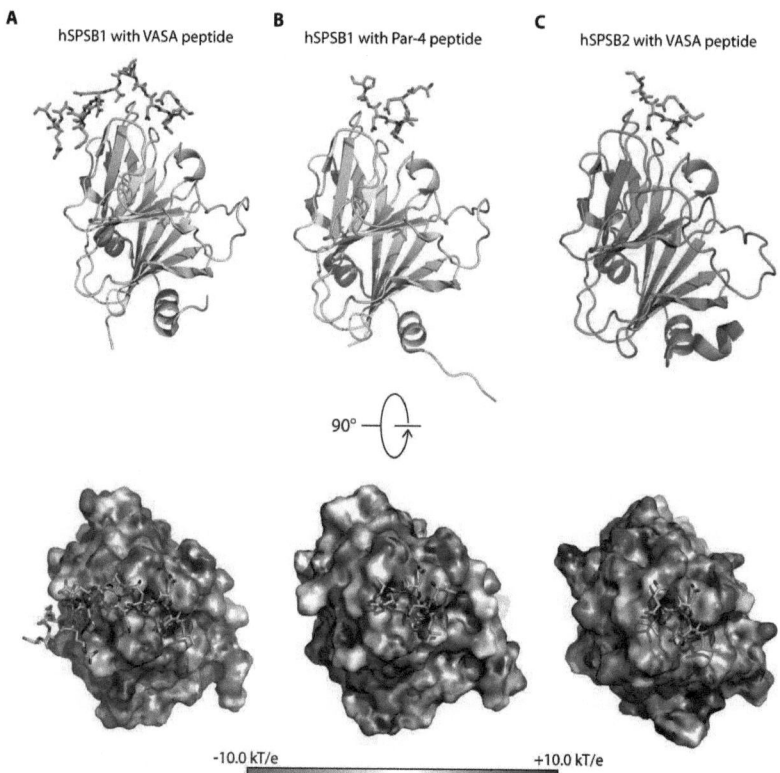

Figure 1.10 **Crystal structures of SPRY domains from SOCS box proteins.** Structures of human SPSB proteins in complex with short peptides, derived from their interactions partners are shown[105]. The structures reveal five extended loops that form an interaction surface, designated surface A[106]. Below the electrostatic surface is shown, that is conserved around the interaction site. PDB codes are as follows: (A) - hSPSB1/Par4 (2JK9), (B) - hSPSB1/VASA (3F2O), (C) - hSPSB2/VASA (3EMW).

Recently, the structure of the SPRY domain of Ash2L, a regulating factor of Histone-methylation has been reported[107]. In Ash2L-SPRY a positively charged surface patch (located similarly as surface A) has been suggested to be the region mediating interaction with other components of the methyltransferase complex. This was substantiated by mutational analysis[107].

For the SPRY domain of DDX1 no information concerning a potential interaction surface is available so far. As DDX1 is commonly found in association with other proteins[108-110], the SPRY domain might be a platform for recruitment and interaction with those factors (**Figure 1.11**). Moreover the SPRY domain is

Introduction

a unique feature of the DDX1 RNA-helicase and its functional relevance for the DEAD-box protein still awaits clarification.

1.3.2 DDX1 in RNA maturation processes

DDX1 is directly or indirectly (via interaction with other factors) involved in various aspects of RNA processing, as shown by several studies[109-114] (**suppl. Table 7.6.1**). Accumulating evidence specifically points towards a role in mRNA maturation as DDX1 is bound to cleavage stimulation factor (CstF-64)[112], interacts with heterogeneous nuclear ribonucleoprotein K (hnRNP K)[109] and binds poly(A) RNA[109]. Moreover it plays a role in mRNA transport [115] and was found in an RNP splicing complex[116]. Recently it was shown that DDX1 is also involved in AU-rich element mediated decay (AMD)[117]. In addition to mRNA maturation, further studies have reported roles of DDX1 in transcription regulation[110] and DNA repair[111]. Some of the studies on DDX1 could map the individual domains, essential for interaction with RNA processing factors (**suppl. Figure 7.6.1**).

Most importantly, DDX1 is found in a multimeric complex in the nucleus and in mRNA transporting granules[115] (**Figure 1.11**). Together with the protein HSPC117 and three additional factors, DDX1 forms the HSPC117 complex[118], which acts as a tRNA ligase[114]. HSPC117 seems to be the major catalytic subunit in this pentameric complex. The three remaining components – Fam98, CGI-99 and Ashwin(=ASW) – of the complex have been identified in pulldown experiments from HEK 293 and HeLa cells[108, 113], but their functions have remained largely uncharacterized[119] (**Figure 1.11**).

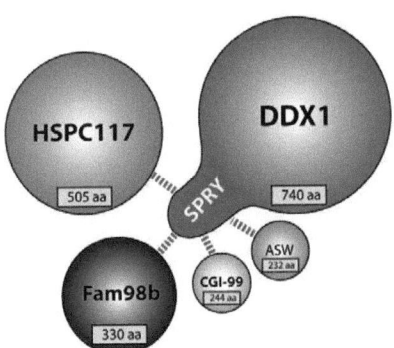

Figure 1.11 **HSPC 117 complex components.** The five different proteins that form the HSPC117 complex are depicted schematically and arranged in a potential complex, stabilized by DDX1. Size of the proteins is indicated by the number of amino acids below. For Fam98b two different isoforms exist, but the 330 amino acid long sequence has been defined as the canonical one.

Although addition of cruciform DNA duplexes that specifically bind to DDX1 inhibited interstrand ligation and tRNA maturation reactions[114], the functional role of the helicase activity of DDX1 in the HSPC117 complex catalyzed ligation process is unknown[119]. Interestingly, efficient depletion of DDX1 did not compromise tRNA maturation as severely as depletion of HSPC117 did[114]. The intriguing question is how an RNA helicase activity could be important for the tRNA ligation reaction. It may partially unwind tRNA to facilitate ligation or simply tether the tRNA to the complex[119]. Depending on DDX1's processivity it could clamp on the tRNA substrate and act as an assembly platform, thereby serving as a structural scaffold for the HSPC117 complex[19].

1.3.3 DDX1's medical relevance

The first study describing DDX1 reported its overexpression in retinoblastoma cell lines[120]. Soon elevated expression levels were also detected in neuroblastoma cell lines[121]. Since then expression of DDX1 in several additional tumor tissues, from germ cell tumors[122] to rhabdomyosarcomas[123] and coamplification with the proto-oncogene MYCN[91, 121] has been reported (**suppl. Table 7.6.2**). In human breast cancer, elevated cytoplasmic DDX1 protein levels were associated with relapse and poor survival[124]. Due to this widespread expression in tumor tissue, DDX1 was recently proposed as a potential biomarker for cancer[125, 126].

In addition to its role in cancer progression, DDX1 is an important host factor for viral infection. This is most probably due to DDX1's central function in RNA processing and therefore the viruses hijack this host protein[127, 128]. Replication of four different virus families – JC virus[129, 130], coronavirus[131], hepatitis C virus[132] and HIV-1[93, 133, 134] – depends on DDX1[128] (**suppl. Table 7.6.3**). Especially the involvement of DDX1 in the export of unspliced HIV-1 mRNAs via the Rev protein has been investigated extensively[127, 135, 136] (**suppl. Figure 7.22**).

Surprisingly despite its involvement as a host factor for viral replication, DDX1 was also reported to serve in the innate immune system as a sensor of viral dsRNA, which is bound directly to RecA-like domain 1[137]. The situation that DDX1 serves in viral defense, but is also hijacked by viruses for their replication sounds a bit paradox, however, this might reflect that DDX1 is a highly contested target in the 'arms race' between the host immune system and viruses[138].

Introduction

1.4 Aims of this thesis

Despite the "birth of the D-E-A-D box" more than 20 years ago[17] and a profound structural characterization of the canonical DEAD-box helicase core in recent years[31, 36], mechanistic and regulatory information is still limited. This situation is further complicated by the different individual activities of each family member that even seem to use different regulatory mechanisms[37]. Moreover, the structural information is limited to either one of the separated RecA-like domains[40, 45, 65] or the minimal helicase fold[31, 33]. Only few structures of auxilliary domains have been reported[35, 41]. These additional domains constitute N- or C-terminal extensions and are separated from the core RecA-like fold. Mechanistic studies, complementing the structures, are sparse and have often relied on short DEAD-box proteins of prokaryotic origin or from lower complexity organisms where molecular standard tools are available[57, 59].

The DEAD-box protein DDX1 harbors the SPRY domain in the helicase core in between the conserved helicase motifs, which suggests a profound structural and functional significance. Therefore, the first aim of this thesis was

> to obtain structural information on the SPRY domain, which is the unique feature of DDX1

The 3-dimensional structure of the SPRY domain was determined at near-atomic resolution by using X-ray crystallography. This approach was complicated by the need for protein crystals that diffract X-rays to high resolution, which required extensive construct optimization and screening. The SPRY structure revealed a compact β-sandwich fold with a conserved patch of positive charge that could serve as interaction platform. The architecture, conservation and functional implications of the SPRY domain are presented in the first results section of this thesis.

Out of the human DEAD-box helicases, DDX1 is the protein which has a special function in RNA maturation[95, 96, 109, 112] and due to its involvement in this fundamental process it also plays a role in cancer progression[94] and viral replication[93, 131, 132]. Mechanistic insights in this protein are therefore of general interest and could be used as a fundament for the design of inhibitors with putative medical relevance. Hence the second aim of this thesis was

to investigate the functional characteristics of DDX1 and to additionally foster our understanding of DEAD-box helicases

The helicase DDX1 protein was studied by a sophisticated combination of biochemical and biophysical approaches. Cooperative effects of ATP/ADP and RNA binding and enzymatic characteristics were investigated. This detailed analysis revealed unique mechanistic properties in comparison to related enzymes and allowed to develop a working model for DDX1. The functional characterization of DDX1 is presented in the second results section of this thesis.

2. Material and Methods

2.1 Materials

2.1.1 Chemicals

The chemicals used in this study were acquired from the following companies:

company	compound
BIOLOG (Bremen)	mant-dADP, mantADP, mant-dATP, mantATP
Bio & Sell (Nürnberg)	agarose, dNTPs
GERBU Biotechnik (Gaiberg)	ampicillin, chloramphenicol, dithioerythritol (DTE), ethylenediaminotetraacetate (EDTA), glycerol, isopropyl β-D-1-thiogalactopyranoside (IPTG), kanamycin, sodium-dodecyl-sulfate (SDS)
GibcoBRL (Karlsruhe)	ethidium bromide solution, peptone, yeast extract
Invitrogen (Karlsruhe)	Topoisomerase kit
Merck (Darmstadt)	ammonium-sulfate, magnesium chloride, potassium dihydrogen phosphate, potassium hydrogen phosphate, potassium chloride, boric acid
Promega (Mannheim)	Pfu-turbo polymerase
New England Biolabs (NEB, Ipswitch)	restriction enzymes, dNTP mix, magnesium sulfate, Vent Polymerase, BSA
Jena Bioscience (Jena)	ATP, ADP, AppNHp, ATPγS
Qiagen (Hilden)	gel-extraction kit, PCR-purification kit, mini-prep kit
Roche Diagnostics (Mannheim)	alkaline phosphatase, nicotinamide adenine dinucleotide (NADH), phosphoenolpyruvate (PEP), pre-mixed lactate dehydrogenase (LDH) and pyruvate kinase (PK) from rabbit muscle (in aqueous glycerol), protease inhibitor tablets, lactate
Roth (Karlsruhe)	ammonium persulfate (APS), isopropanol, 30% (w/v) Acryl amide : 0,8% (w/v) Bis-acryl amide stock solution (37.5:1)
Serva (Heidelberg)	Bromphenol blue, BSA, Coomassie Brilliant Blue R/G-250
Expedeon (Harston, UK)	InstantBlue protein stain
Sigma/Fluka (Deisenhofen)	acetic acid, phenylmethanesulfonylfluoride (PMSF), tetramethylethylenediamine (TEMED), Thrombin, Trizma (Tris), 4-(2-hydroxyethyl)-1-piperazineethanesulfonic acid (HEPES),
Sigma Aldrich (Steinheim)	ß-mercaptoethanol (β-ME), p-nitrophenyl α-glucopyranoside, p-Aminobenzamidine dihydrochloride, ethanol, imidazole, tris(2-carboxyethyl)phosphine (TCEP)

In the experiments double-distilled and de-ionized ELGA-water (purchased from VEOLIA water, Celle) was used for buffer- and sample preparation.

2.1.2 Buffers

buffer denomination	composition
Lysis buffer A	50 mM Tris-HCl pH 8.0
	250 mM KCl
	10 mM β-ME
Elution buffer A	50 mM Tris-HCl pH 8.0
	200 mM KCl
	10 mM β-ME
	250 mM Imidazole
buffer A_{hep}	50 mM Tris-HCl pH 8.0
	5 mM $MgCl_2$
	3 mM DTE
buffer B_{hep}	50 mM Tris-HCl pH 8.0
	5 mM $MgCl_2$
	3 mM DTE
	1 M KCl
buffer A_{monoQ}	50 mM Tris-HCl pH 9.0
	5 mM $MgCl_2$
	3 mM DTE
buffer B_{monoQ}	50 mM Tris-HCl pH 9.0
	5 mM $MgCl_2$
	3 mM DTE
	1 M KCl
buffer A_{monoS}	50 mM MES-NaOH pH 5.5
	5 mM $MgCl_2$
	3 mM DTE
buffer B_{monoS}	50 mM MES-NaOH pH 5.5
	5 mM $MgCl_2$
	3 mM DTE
	1M KCl
storage buffer	10 mM HEPES-NaOH pH 8.0
	250 mM KCl
	5 mM $MgCl_2$
	3 mM DTE
EMSA reaction buffer	50 mM Tris-HCl pH 8.0
	200 mM KCl
	10 mM $MgCl_2$.
oligonucleotide annealing buffer	20 mM Tris-HCl pH 7.5
	200 mM K-Acetate
	0.1 mM EDTA

Materials and Methods

4 x unwinding buffer	100 mM Tris-HCl pH 8.0
	400 mM KCl
	10 mM DTE
RNA loading dye	40 % (v/v) glycerol
	(50 mM EDTA pH 7.5)
	0.1 mg bromphenol blue
2 x ATPase buffer	50 mM Tris-HCl pH 8.0
	1 mM EDTA
	250 mM KCl
	0.5 mM NADH
	0.8 mM PEP
	2.8 units/ml PK
	4 units/ml LDH
	2 mM $MgCl_2$
titration buffer	50 mM HEPES-NaOH pH 8.0
	5 mM $MgCl_2$
	250 mM KCl
5 x SDS-loading buffer	50 mM Tris-HCl pH 7.0
	10% (w/v) SDS
	0.1% (w/v) bromphenol blue
	1% (v/v) β- ME
	14% (w/v) DTE
	10% (v/v) glycerol
6 x DNA-loading buffer	30% (v/v) glycerol
	0.25% (w/v) bromphenol blue
	0.25% (w/v) xylene cyanol FF
SDS running buffer	25 mM Tris-HCl pH 8.0
	200 mM glycine
	0.1% (w/v) SDS
running gel buffer	1.5 M Tris-HCl pH 8.8
	0.4% (w/v) SDS
stacking gel buffer	0.5 M Tris-HCl pH 6.8
	0.8 % (w/v) SDS
TB/competence buffer	10 mM PIPES-HCl pH 6.7
	55 m M $MnCl_2$
	15 mM $CaCl_2$
	250 mM KCl
	7 % (v/v) DMSO
TBE	89 mM Tris-HCl pH 8.0
	89 mM boric acid
	2 mM EDTA

Most buffers and stock-solutions, if not containing DTE or β-ME were stored at 297 K for several days. Buffer stock solutions of BSA, MOPS, NADH and PEP were stored at 253 K. Chromatography-, lysis- and assay-buffers were prepared fresh for each experiment.

2.1.3 Software

The following software was used during the study for data acquisition, handling, evaluation and experimental design:

company	software
Adobe (München)	Adobe CS 4.0, Adobe Illustrator, Acrobat 8.0 professional
Biokin ltd. (Watertown, USA)	Dynafit 4.05
Bioinformatics.org (USA)	PrimerX
Barton, G.J. - *Prot. engineering* (1993)	ALscript[139]
Collaborative Computational Project 4 (Oxon, UK)	CCP4 software suit
DeLano Scientific LLC (USA)	PyMOL
dotPDN LLC (USA)	paint.net v 3.36
Epson (Meerbusch)	Epson Scan
Erithacus software (Horley, UK)	GraFit 3.0
European Bioinformatics Institute (Cambridge, GB)	ClustalW 2.0.1, ClustalOmega
GraphPad Software (San Diego, USA)	Prism 5.01
Jasco (Groß-Umstadt)	Spectra manager
Livingstone and Barton - *CABIOS* (1993)	AMAS server[140]
Microsoft (Unterschleißheim)	office-suite professional 2003
Sci-Ed (Cary, USA)	Clone-manager 9.0
Swiss Institute of Bioinformatics (CH)	ExPASy proteomics server
Wolfgang Kabsch (MPI for medical research)	XDS, X-ray detector software
zbio.net/Stormoff (Düsseldorf)	six-frame translation

Note that all simple sketch-like figures in this study were created with the software paint.net (Rick Brewster, dotPDN LLC, USA). Schematic figures were created with Adobe Illustrator (Adobe, München, Germany).

All data representations were created with Prism 5.01 (GraphPad Software, San Diego, USA) or with GraFit 3.0 (Erithacus software, Sussex, UK). The fits shown were either created with Prism or GraFit. Fits created with Dynafit were plotted in Prism.

Multiple sequence alignments were createad with either ClustalOmega or ClustalW from the European Bioinformatics Institute of the EMBL (EMBL-EBI, www.ebi.ac.uk/tools/msa). Secondary structure predictions were obtained with PSIPRED from the Bloomsbury Centre for Bioinformatics (bioinf.cs.ucl.ac.uk/psipred). Visualisation of sequence alignments was realized with the Linux software ALscript[139] and sequence conservation values were obtained with the AMAS-server[140].

Materials and Methods

2.1.4 Crystallization screens

The following screens were used for crystal screening in 96-well formate:

company	screen
Qiagen (Germany)	classics suite I and II
Qiagen (Germany)	Joint Consortium for Structural Genomics (JCSG) – core suite
Qiagen (Germany)	PEG I and II suites
Emerald (USA)	Wizard I and II
Hampton Research (USA)	additive screen

2.1.5 Growth media

Bacteria were grown in lysogeny broth/Luria Bertani (LB) medium, supplemented with appropriate antibiotics. To obtain solid LB-agar plates, medium was supplemented with 15 g/l bacto-agar. All media and plates were prepared and autoclaved by the media kitchen of the Max Planck Institute for Medical Research (Heidelberg).

LB medium	10 g/l bacto-tryptone
	5 g/l bacto-yeast extract
	10 g/l NaCl
	(adjust pH with 10 M NaOH to 7.0)
LB agar	15 g/l bacto agar
	25 g/l standard-nutrient-broth

Antibiotics were added to the media according to the resistance cassettes integrated into the chromosomes of the respective bacteria strains and according to the resistance marker on the plasmid DNA:

vector	antibiotic
pET28a	50 µg/ml kanamycin
pGEX-4T-1	100 µg/ml ampicillin
pET22b	100 µg/ml ampicillin
pET21d	100 µg/ml ampicillin
pRIL	34 µg/ml chloramphenicol

For selenomethionine labeling of target proteins, *E. coli* strain BL21-CodonPlus(DE3)-RIL cells were grown in minimal media supplemented with selenomethionine instead of methionine. The media were prepared according to Van Duyne et al., 1991[141].

2.1.6 Bacterial strains

The *E.coli* strain DH5α was used for cloning of plasmid DNA, whereas the BL21-CodonPlus(DE3)-RIL cells were used for protein expression. For some expression trials also the strain K12 ER2508 was used, which is a derivative of RR1 that has been made *lon*[-142]. All cells were stored at 193 K in TB/competence-buffer.

strain	genotype	source
DH5α	F- φ80*lacZ*ΔM15 Δ(*lacZYA-argF*) U169 *recA1 endA1 hsdR17* (rk-, mk+) *phoA supE44 thi-1 gyrA96 relA1* λ-	Stratagene (Cedar Creek, TX, USA) [143]
BL21-CodonPlus(DE3)-RIL	ompT, hsdSB (rB-, mB-), gal(λcIts857,ind1,Sam7, nin5, lacUV5-T7gene1), dcm(DE3)	Invitrogen (USA) [144]
K12 ER2508	F- *ara-14 leuB6 fhuA2* Δ(*argF-lac*)*U169 lacY1 lon::miniTn*10(Tet^R) *glnV44 galK2 rpsL20*(Str^R) *xyl-5 mtl-5* Δ(*malB*) *zjc::Tn5*(Kan^R) Δ(*mcrC-mrr*)$_{HB101}$	NEB (Ipswich, UK) [142, 145]

2.2 Molecular biology

2.2.1 Transformation protocol for chemically competent *E. coli* cells

Chemically competent cells were prepared according to a protocol from Hanahan et al. [146] and were kindly provided by Maike Gebhardt, Christina Sendlmeier and Florence Jungblut (MPI for medical research). For plasmid tranformation 25 µl aliquots of respective *E. coli* cell suspension was thawed on ice and 1 µl plasmid DNA (10 – 50 ng/µl) was added carefully. Cells were incubated with the plasmid DNA for ~15 min on ice, „heat-shocked" at 315 K for 20 sec and afterwards immediately placed on ice. After 3 min incubation on ice, 450 µl pre-warmed (310 K) LB medium was added to the cells and they were incubated at 310 K, shaking at 750 rpm in a thermomixer comfort (Eppendorf, Hamburg, Germany) for 1 h. Cells were harvested by centrifugation at 4500 rpm (2.400 x g) in a table-top-centrifuge (Heraeus® Biofuge®, Thermo-Scientific, Waltham, USA). Supernatant was discarded and cells were resuspended in 100 µl of fresh LB medium and plated on a pre-warmed LB-agar plate containing antibiotics according to the respective resistance cassette on the plasmid. Plates were incubated at 310 K for 12 h or until single bacteria colonies were visible.

Materials and Methods

2.2.2 Amplification of DNA fragments using polymerase chain reaction (PCR)

cDNA was used as a PCR-template to amplify genomic fragments corresponding to open-reading-frames (ORFs). PCR primers were designed to contain restriction sites and ordered in the form of chemically synthesized oligonucleotides from MWG-eurofins (Ebersberg, Germany). Thermocycling was conducted following a three-step protocol[147] (**see Thermocycling protocol below**) in a TPersonal machine (Biometra, Göttingen, Germany). PCR reagents were purchased from New England Biolabs (NEB, Schwalbach, Germany) and parameters such as annealing temperature and elongation time were adapted to product length and primer properties.

Reagent	Concentration	Volume
10 x Thermopol buffer*		5 µl
template cDNA	~ 50 ng/µl	0.5 µl
dNTP's	10 mM	5 µl
Forward primer	0.1 mM	0.5 µl
Reverse primer	0.1 mM	0.5 µl
Formamid	> 99.5 %	3.5 µl
MgSO$_4$	100 mM	0-2 µl
Vent polymerase	2 U/µl	0.5 µl
ddH$_2$O	> 99.9 %	add to 50 µl

* 200 mM Tris-HCl pH 8.8, 100 mM (NH$_4$)SO$_4$, 100 mM KCl, 20 mM MgSO$_4$, 1 % (v/v) Triton X-100

Thermocycling protocol

Reaction step	Temperature [K]	Time [sec]	
Initial denaturation	365	120	
Melting	365	30	
Annealing	327	30	x 30
Elongation	345	90 (60 per kb)	
Final elongation	345	600	
Storage	277	∞	

PCR products were separated by agarose gel electrophoresis (**see section 2.2.6**). Separated DNA fragments were visualized by illuminating the gel with UV-light at 280 nm (excitation of intercalated dye), fragments were excised and DNA was purified by using a gel-extraction kit, following the manufacturer's protocol (Qiagen, Hilden, Germany).

2.2.3 Site directed mutagenesis

Simple mutations (like e.g. single codon changes or amino acid substitutions) were introduced in plasmid templates by using site-directed mutagenesis (SDM) according to the QuikChange® protocol

(Stratagene, La Jolla, USA). Primers suitable for SDM reactions were designed with the online software PrimerX[148] and chemically synthesized by MWG-eurofins (Ebersberg, Germany). QuikChange® thermocycling was conducted in a TPersonal machine (Biometra, Göttingen, Germany).

Reagent	Concentration	Volume
10 x Pfu buffer*		5 µl
template plasmid	~ 50 ng/µl	1 µl
dNTP's	10 mM	1 µl
Forward primer	0.01 mM	1.25 µl
Reverse primer	0.01 mM	1.25 µl
Pfu Turbo polymerase	2 U/µl	1 µl
ddH$_2$O	> 99.9 %	add to 50 µl

* 200 mM Tris-HCl pH 8.8, 100 mM (NH$_4$)SO$_4$, 100 mM KCl, 1 % (v/v) Triton X-100, 1 mg/ml BSA

SDM QuikChange® thermocycling protocol

Reaction step	Temperature [K]	Time [sec]	
Initial denaturation	368	30	
Melting	368	30	
Annealing	328	60	x 16
Elongation	341	480 (60 per kb)	
Final elongation	341	600	
Storage	277	∞	

After DNA amplification (by thermocycling), the reaction products were digested with DpnI (10 U) for 2 h at 310 K to remove all parental (methylated) DNA. Then, the nicked PCR product was transformed into E. coli DH5α cells. Successful mutagenesis was verified by isolating plasmid DNA from single bacteria colonies and next-generation DNA sequencing (done by MWG-eurofins, Ebersberg, Germany).

2.2.4 Preparation of plasmid DNA

Plasmid DNA was prepared from single colonies of E. coli DH5α cells that grew on LB plates after transformation and plating. Single colonies were used to inoculate 5 ml LB culture, containing the appropriate antibiotics and cultures were grown at 310 K, shaking at 105 rpm for 14 h. Bacteria were harvested by centrifugation at 4.000 rpm (3.220 x g) in a 5810 R centrifuge (Eppendorf, Hamburg, Germany). Plasmid DNA was isolated from the cell pellet according to the QIAprep mini-prep protocol using the Spin Miniprep kit (Qiagen, Hilden, Germany). Final concentrations were determined spectroscopically by measuring absorption at a wavelength of 260 nm on a NanoDrop ND-1000 spectrophotometer (NanoDrop Technologies, Wilmington, USA).

2.2.5 Digest by restriction enzymes and ligation of nucleotide fragments

Restriction endonucleases were used to digest PCR products and to generate sticky or blunt ends for ligation of nucleotide fragments in the multiple-cloning-site (MCS) of the respective target vector. Plasmid DNA was incubated with restriction enzymes (purchased from NEB, Schwalbach, Germany) for 2 – 12 h at 310 K, depending on the respective endonuclease activity. After the restriction digest, samples were purified either via agarose-gel-electrophoresis (**see section 2.2.6**) or by using the PCR purification kit (Qiagen, Hilden, Germany), depending on fragment size.

Reagent	Concentration	Volume
10 x NEB buffer*		5 µl
Bovine Serum Albumin (BSA)*	10 mg/µl	0.5 µl
Restriction enzyme 1	10 – 20 U/µl	1 µl
Restriction enzyme 2	10 – 20 U/µl	1 µl
PCR product/plasmid DNA	~ 50 ng/µl	10 – 40 µl
ddH$_2$O	> 99.9 %	add to 50 µl

* Depending on Restriction enzyme used, either NEB buffer 1,2,3 or 4 with or without BSA was used

The resulting nucleotide fragments from the restriction reaction, i.e. digested vector and insert with compatible sticky ends, were ligated. Ligation was conducted according to the manufacturer's protocol using the rapid DNA ligation kit (Fermentas, St. Leon-Rot, Germany). Correct concentrations of vector and insert for the ligation reaction (to achive a 1:3 ratio) were calculated using the ligation calculator of the university of Düsseldorf (insilico.uni-duesseldorf.de/Lig_Input).

Reagent	Concentration	Volume
5 x rapid ligation buffer[a]		4 µl
insert[b]	25 – 50 ng	5-14 µl
vector[b]	25 – 50 ng	5-14 µl
T4 DNA ligase	10 U/µl	1 µl
ddH$_2$O	> 99.9 %	add to 20 µl

[a] 400 mM Tris-HCl, 100 mM MgCl$_2$, 100 mM dithiothreitol (DTT), 5 mM ATP
[b] vector and insert concentrations were calculated individually to obtain a vector to insert ratio of 1:3

Ligation reactions were incubated at 297 K for 10 min and then transformed in *E. coli* DH5α cells.

2.2.6 Agarose gel electrophoresis

Agarose gel electrophoresis was used to separate nucleotide fragments, based on their electrophoretic mobility. To prepare gels 0.5 – 1.5 % (w/v) agarose (Bio & Sell, Feucht, Germany) was suspended in 1 x TBE buffer, heated to 368 K and poured in a plastic-tray before solidifying upon cooling down. Agarose gels were supplemented with 0.003 % (v/v) gel-red (Roth, Karlsruhe, Germany) to stain nucleotide strands. Samples were supplemented with a loading-dye, to increase viscosity before loading. To estimate DNA size 1 µl of a NEB ruler, 1 kb DNA ladder (NEB, Schwalbach, Germany) was loaded. Electrophoresis was performed using a PowerPac™ High-Voltage power supply (Bio-Rad, Munich, Germany) at 100 V for 1-2 h. Separated DNA fragments were visualized by exciting intercalated dye at 302 nm. Fragments were excised and DNA was purified by using the gel-extraction kit, following the manufacturer's protocol (Qiagen, Hilden, Germany).

2.2.7 Sequencing

Successful cloning was verified by test digests only in situations, where restriction fragments helped to identify the presence of an insert or changed upon incorporation of a mutation. Constructs, which showed a correct restriction pattern in test digests or QuikChange® SDM products were further checked by sequencing. All sequencing reactions were performed by MWG-eurofins (Ebersberg, Germany) on ABI 3730XL sequencing machines (Life Technologies GmbH, Darmstadt, Germany) and evaluated by MWG-eurofins (Ebersberg, Germany). The material required for a typical sequencing reaction was 14 µl DNA sample of a concentration of 5 – 300 ng/µl mixed with 1 µl sequencing primer (stock 0.01 mM). Material for sequencing reactions was sent to MWG-eurofins (Ebersberg, Germany) via letter mail.

Note that all cloning and sequencing primer are listed in the appendix

(**section 7.7**).

2.2.8 Cloning of the DDX1 ORF into pET28a

The coding sequence of human DDX1 was amplified from cDNA (Accession: BC012739, clone ID: 3835131) obtained from OpenBiosystems (Schwerte, Germany) by PCR and cloned into pET28a expression vector DNA (Novagen, Darmstadt, Germany). Briefly, a colony PCR was performed on bacteria containing the cDNA vector with the DDX1 ORF. Primers did contain restriction sites for cloning of the amplified DDX1-PCR fragment in the target vector. The forward primer (DDX1_NheI_fwd) was designed to contain a *Nhe*I-restriction-site upstream of the ORF and the reverse primer (DDX1_NotI_rev) was designed to contain a *Not*I-restriction-site downstream of the ORF. After DNA amplification, the PCR reaction was separated by agarose gel-electrophoresis and the band corresponding to the DDX1 DNA fragment (according to electrophoretic mobility) was excised and purified with a gel-extraction kit

Materials and Methods

(Qiagen, Hilden, Germany). The DDX1 PCR product DNA and the pET28a vector DNA were both digested with *Nhe*I- and *Not*I-restriction-endonucleases at 310 K for 12 h. After the digest, the linearized plasmid DNA and the DDX1 PCR product DNA were separated via agarose gel-electrophoresis, the corresponding bands were excised and the DNA was extracted from the gel slices via the gel-extraction kit (Qiagen, Hilden, Germany). The DDX1 PCR product DNA was ligated with the linearized pET28a vector DNA using the rapid DNA ligation kit (Fermentas, St. Leon-Rot, Germany). Ligation in pET28a resulted in a DDX1 construct (= pET28a(DDX1-WT)) harboring the sequence for a fused N-terminal hexa-histidine tag (6xHis-Tag). The N-terminal 6xHis-Tag is fused to the DDX1 ORF via a Thrombin cleavage site.

2.2.9 Cloning of the DDX1 ORF into pET22b

To test for the effects of protein expression in the periplasm, the sequence encoding for DDX1 was subcloned into pET22b vector DNA. This vector carries an N-terminal *pelB* signal sequence for periplasmic protein localization[149]. The sequence, coding for the entire DDX1 ORF and the N-terminal 6xHis-Tag, was cloned from pET28a(DDX1-WT) as *Nco*I – *Xho*I restriction fragment into pET22b vector DNA. Therefore, vector DNA from both, pET28a(DDX-WT) and pET22b was digested with *Nco*I- and *Xho*I-restriction-endonucleases at 310 K for 2 h. Digestion products were separated via agarose gel-electrophoresis, excised from the gel and purified via the gel-extraction kit (Qiagen, Hilden, Germany). The DDX1 digestion product was ligated with the linearized pET22b vector DNA using the rapid DNA ligation kit (Fermentas, St. Leon-Rot, Germany). Ligation resulted in a DDX1 construc (= pET22b(DDX1-WT))t harboring the codon triplets for an N-terminal *pelB* signal sequence, followed by a 6xHis-Tag.

2.2.10 Cloning of the DDX1 ORF into pGEX-4T-1

To test for the effects of a GST-tag fusion on protein stability, DDX1 was subcloned into pGEX-4T-1 vector DNA. This vector carries the sequence for an N-terminal GST-Tag to facilitate protein folding. SDM was used to introduce a *Bam*HI-restriction-site upstream of the DDX1 ORF in pET28a (primers DDX1_genBamHI-pGEX_f and _r, for complete sequence **see appendix section 7.4**). The DDX1 ORF was cloned as a *Bam*HI – *Not*I restriction fragment from pET28a vector DNA into pGEX-4T-1 vector DNA. Cloning and ligation were conducted as mentioned above resulting in pGEX-4T-1(DDX1-WT).

2.2.11 Cloning of C-terminally truncated DDX1 variants in pET28a

Several C-terminally truncated variants of DDX1 were constructed in vector pET28a, using pET28a(DDX1-WT) as a template. QuikChange® SDM was used to introduce double stop-codons (sequence TGA TAA) at desired positions within the DDX1 ORF. Using the appropriate SDM-primers (**see appendix section 7.4**, SDM primer table) the following truncated DDX1-variants were constructed: pET28a(DDX1-432); pET28a(DDX1-610); pET28a(DDX1-648); pET28a(DDX1-655); pET28a(DDX1-674); pET28a(DDX1-694); pET28a(DDX1-728) - the number denotes the position of the C-terminal truncation (also **see table 3.1** in the results part).

2.2.12 Cloning of DDX1ΔSPRY in pET28a

The sequence, coding for the entire SPRY domain was removed from pET28a(DDX1-WT), resulting in construct pET28a(DDX1ΔSPRY). For this, endonuclease restriction sites were introduced upstream and downstream of the SPRY domain coding sequence via QuikChange® SDM. A *Bam*HI-restriction-site was introduced after the codon triplett for lysine residue 69 and a second *Bam*HI-restrction-site was introduced before the codon triplett for lysine residue 248 in pET28a(DDX1-WT) vector DNA. The DNA of the mutated construct was digested with *Bam*HI-restriction-endonuclease, the reaction products were separated via agarose gel-electrophoresis and the vector DNA lacking the SPRY domain sequence was re-ligated.

2.2.13 SDM to generate a Walker A Lysine mutant of DDX1 in pET28a

The codon triplett for the essential Walker A lysine (GSG<u>K</u>T motif) at position 52 was mutated to alanine via QuikChange® SDM, resulting in construct pET28a(DDX1-K52A). To ensure correct mutation the plasmid was sequenced and the altered residue was additional verified by ESI-MS on the protein level (**see below section 2.3.7**), conducted by Melanie Müller and Marion Gradl.

2.2.14 Cloning of the SPRY domain in pET28a

In pET28a(DDX1-WT) a *Bam*HI-restriction-site was introduced upstream (primers DX1SPRY_1gBamH-f and –r) and an *Xho*I-restriction-site downstream (primers DX1SPRY_2gBamH-f and –r, for complete sequence **see appendix section 7.4**) of the sequence, coding for the separated SPRY-domain (residues 72-283 on amino acid level) via QuikChange® SDM. The mutated vector DNA was then digested with *Bam*HI- and *Xho*I-restriction-endonucleases and the excised DNA encoding for the SPRY domain was ligated with pre-linearized pET28a vector DNA. The resulting construct was designated pET28a(SPRY_72-283).

Additional N- and C-terminally truncated SPRY variants were generated using this construct as a template: pET28a(SPRY_84-283); pET28a(SPRY_100-283); pET28a(SPRY_72-261); pET28a(SPRY_84-261); pET28a(SPRY_100-261) - constructs are summarized in **table 2.1**.

Table 2.1 Constructs of the DDX1-SPRY domain

construct	length (amino acids, referring to DDX1)	vector	Cloning strategy
SPRY_72-283	Glu72 - Lys283	pET28a	BamHI – XhoI restriction fragment from pET28a(DDX1)
SPRY_84-283	Gly84 - Lys283	pET28a	Removed N-terminal BamHI – BamHI restriction fragment from pET28a(SPRY_72-283)
SPRY_100-283	Ser100 - Lys283	pET28a	Removed N-terminal BamHI – BamHI restriction fragment from pET28a(SPRY_72-283)
SPRY_72-261	Glu72 - Ala261	pET28a	Introduced Stop-codon at the sequence coding for the C-terminus of pET28a(SPRY_72-283)
SPRY_84-261	Gly84 - Ala261	pET28a	Introduced Stop-codon at the sequence coding for the C-terminus of pET28a(SPRY_84-283)
SPRY_100-261	Ser100 - Ala261	pET28a	Introduced Stop-codon at the sequence coding for the C-terminus of pET28a(SPRY_100-283)

The N-terminal truncated SPRY variants were obtained by introducing BamHI-restriction-sites upstream and downstream of the region to be removed by SDM. Constructs were digested with BamHI-restriction-endonuclease and the vector DNA lacking the BamHI – BamHI restriction fragment was re-ligated. C-terminal truncated SPRY variants were obtained by introducing Stop-codons via SDM.

2.2.15 Cloning of subunits of the HSPC117 complex into pET28a

The components of the HSPC117 complex were cloned either from cDNA or from other vector DNA into pET28a expression vector DNA (Novagen, Darmstadt, Germany). The coding sequence for the HSPC117 proteins Fam98b and CGI-99 was available in the pFBDM vector DNA (cloned by Maike Gehardt) and was amplified via PCR, using the pFBDM vector DNA as a template. The coding sequence of both proteins was amplified with primers coding for a NheI-restriction-site upstream and an XhoI-restriction-site downstream of the respective protein ORF. PCR products were then purified, digested with NheI- and XhoI-restriction-endonucleases and ligated with pre-linearized pET28a vector DNA, resulting in the expression constructs pET28a(Fam98b) and pET28a(CGI-99).

The coding sequence for the protein ASW/C2ORF49 was amplified from cDNA (Accession: BC001310 Clone ID: 3453623) obtained from OpenBiosystems (Schwerte, Germany) by PCR and cloned into pET28a expression vector DNA (Novagen, Darmstadt, Germany). The forward primer (ASW_NdeI_fwd2) was designed to contain a NdeI-restriction-site upstream of the ORF and the reverse primer (ASW_BamHI_rev1) was designed to contain a BamHI-restriction-site downstream of the ORF. After DNA amplification by thermocycling, the PCR product was digested with NdeI- and BamHI-restriction-endonucleases and ligated with pre-linearized pET28a vector DNA to yield pET28a(ASW).

2.3 Protein biochemistry

2.3.1 Recombinant protein expression

Proteins were expressed in BL21-CodonPlus(DE3) RIL *E. coli* cells (Invitrogen, Carlsbad, USA) that were transformed with the respective expression plasmid DNA prior to cultivation. Baffled glass flasks with a total volume of 5 l (Schott AG, Mainz, Germany) were filled with 2 l of LB medium each, supplemented with selection antibiotics and inoculated with 5 – 25 ml (per 2 l) of an overnight bacteria pre-culture. Cells were grown at 310 K, shaking in a Multitron incubation shaking device (INFORS HT, Bottmingen, Switzerland) for 2-3 h until an OD_{600} ~ 0.8 was reached. Cultures were cooled down to 291 K and protein expression was induced by addition of 0.5 mM isopropyl-β-D-thiogalactosid (IPTG). Cultures were then incubated shaking at 291 K for ~ 14 h. Bacteria cells were harvested by centrifugation at 6.000 x g in a Sorvall Evolution RC 2925 centrifuge (Thermo Fisher, Waltham, USA) at 277 K for 20 min. Supernatant was discarded and cell pellets were dissolved in ~10 ml lysis buffer A (see section 2.1.2, buffer table) per 2 l of input LB medium and either directly used for protein purification or shock frozen in liquid nitrogen and stored at 193 K.

2.3.2 Analysis of protein expression and purity via SDS- and native PAGE

All protein expression and purification steps were checked by discontinuous sodium-dodecyl-sulfate polyacrylamide gel-electrophoresis (SDS-PAGE)[150]. Gels containing 10 – 17 % (w/v) acrylamide (depending on desired resolution) were cast in the size 8.6 x 6.8 cm (mini-gels) using the Protean 4 system (Bio-Rad, Munich, Germany).

15 % running gel

Component	Volume	Supplier
ddH$_2$O	7.3 ml	ELGA, Celle, Germany
running-gel-buffer*	7.5 ml	see buffer list, section 2.1.2
30 % (w/v) acrylamide-,bisacrylamide (37.5:1)	15 ml	Roth, Karlsruhe, Germany
TEMED	25 µl	Serva, Heidelberg, Germany
added directly before casting to start polymerization reaction		
10% (w/v) APS	250 µl	Merck, Darmstadt, Germany
*running-gel-buffer was replaced	by either 1xTBE or 40 mM CHES	for native gels

To remove air-bubbles the running gel was overlaid with isopropanol. After polymerization of the running-gel, the isopropanol was removed and a stacking-gel (with different pH) was cast on top.

stacking gel

Component	Volume	Supplier
ddH$_2$O	8.9 ml	ELGA, Celle, Germany
stacking-gel-buffer	3.8 ml	see buffer list, section 2.1.2
30% (w/v) acrylamide-,bisacrylamide (37.5:1)	2.3 ml	Roth, Karlsruhe, Germany
Tetramethylethylenediamine (TEMED) *added directly before casting to start polymerization reaction*	15 µl	Serva, Heidelberg, Germany
10% (w/v) APS	150 µl	Merck, Darmstadt, Germany

A 15-well polycarbonate comb (Bio-Rad, Munich, Germany) was inserted into stacking gels before polymerization, resulting in wells of 3.35 mm width. Protein samples were denatured prior to gel-loading, by addition of SDS-loading buffer and heating to 368 K for 5 min. Electrophoresis was performed with PowerPack 200 Voltage-supply (Bio-Rad, Munich, Germany) in SDS-PAGE running buffer at 40 mA per gel for ~ 40 min. Gels were stained using InstantBlue protein stain (Expedeon, Harston, UK).

Helicase and gel-shift assays were done using native PAGE. Native PAGE was performed similar as described for SDS-PAGE, except acrylamide gels were prepared with either 1 x TBE pH 8.3 (for helicase assays) or 40 mM CHES-NaOH pH 10.0 (for gel-shift assays) instead of running-gel-buffer and no stacking gels were used. Furthermore, the 30% (w/v) acrylamide-,bisacrylamide (37.5:1) solution was exchanged for a 40% (w/v) acrylamide-,bisacrylamide (19:1).

Samples were loaded on the gel in 8 % (v/v) glycerol (stained with 0.01 mg bromphenol blue per 100 µl) instead of SDS-loading buffer. For the gel-shift experiments, gels were run with 200 V per gel at 277 K for 60 min. For the helicase experiments, gels were run with 60 V per gel at 277 K for 40 min. Electrophoresis was performed in either 1 x TBE pH 8.3 (for helicase assays) or 40 mM CHES-NaOH pH 10.0 (for gel-shift assays), thermostated to 277 K.

2.3.3 Cell lysis and clearification of lysate

Cell pellets from 4 – 6 l *E. coli* culture in lysis buffer A (see section 2.1.2, buffer list), overexpressing the respective protein, were directly processed after protein expression or thawed from 193 K. Cell walls were disrupted by pulse-sonication (Branson Sonifier, Dietzenbach, Germany) at 277 K for 20 min. Lysate was cleared by ultracentrifugation at 125.000 x g in a L8-70M ultracentrifuge with a 45Ti rotor

(Beckman Coulter, Krefeld, Germany) at 277 K for 40 min. The cleared supernatant was used for purification of recombinant proteins.

2.3.3 Purification of full-length and truncated DDX1 variants

The same purification protocol was used for full-length DDX1, C-terminally truncated DDX1 constructs and other DDX1 mutagenesis variants. The cleared supernatant of bacteria lysate was loaded on a 1 ml Ni^{2+}-NTA HiTrap column (GE Healthcare, Freiburg, Germany), pre-equilibrated with lysis buffer A using a peristaltic pump P-1 (GE Healthcare, Freiburg, Germany) at 277 K. The column was washed with three column volumes (CVs) of lysis buffer A and with one CV of lysis buffer A containing additionally 1 M KCl to remove nucleotides. Bound proteins were eluted in a linear gradient of 28 CVs (40 min gradient at 0.7 ml/min) to elution buffer A, containing 250 mM imidazole using an ÄKTA purifier 10 (GE Healthcare, Freiburg, Germany). After elution from the 1ml Ni^{2+}-NTA HiTrap column (GE Healthcare, Freiburg, Germany), homogeneous fractions of DDX1 were pooled and diluted in buffer A_{hep}. DDX1 was loaded on a 1 ml HiTrap heparin column (GE Healthcare, Freiburg, Germany), equilibrated with buffer A_{hep} using an ÄKTA purifier 10 at 277 K. Proteins were eluted in a linear gradient of 28 CVs to buffer B_{hep} containing 1 M KCl. Elution fractions, containing protein were diluted in buffer A_{monoQ} and loaded on a MonoQ 10/100 GL column (GE Healthcare, Freiburg, Germany) equilibrated with buffer A_{monoQ}. Bound proteins were eluted in a linear gradient of 28 CVs to buffer B_{monoQ}, containing 1 M KCl and protein fractions were concentrated using Amicon Ultra 10K centrifugal filter units (Milipore, Darmstadt, Germany) with a molecular weight cut-off (MWCO) of 10 kDa. Further purification was achieved by gel-filtration chromatography on a Superose 6® 10/300 GL column (GE Healthcare), equilibrated with storage buffer. Fractions containing pure DDX1 protein were pooled, concentrated in Amicon Ultra 10K filters and stored in ~ 20 mg/ml (200-300 µM) aliquots at 193 K.

	column purification steps	
1.)	Ni^{2+}-NTA HiTrap column	Binding of His-tagged proteins
2.)	HiTrap heparin column	Retention of ATP binding proteins, removal of nucleotides
3.)	MonoQ 10/100 GL column	Anion-exchange to retain negatively charged proteins
4.)	Superose 6 10/300 GL column	Polishing by size-exclusion chromatography

Protein purity during all steps was verified by 15 % (w/v) SDS-PAGE. Purified DDX1 protein constructs were confirmed to run as a single protein band corresponding to their respective electrophoretic mobility. Protein identity was further verified in experiments conducted by Melanie Müller and Marion Gradl using mass spectrometry (**see section 2.3.7**).

Materials and Methods

2.3.4 Purification of the SPRY domain of DDX1

The protocol for the purification of the separate SPRY domain was slightly modified from the DDX1 purification protocol. Cleared supernatant of bacteria lysate was loaded on a 1 ml Ni^{2+}-NTA HiTrap column (GE Healthcare, Freiburg, Germany), pre-equilibrated with lysis buffer A, using a peristaltic pump P-1 (GE Healthcare, Freiburg, Germany) at 277 K. The column was washed with three column volumes (CVs) of lysis buffer A and with one CV of lysis buffer A containing additionally 1 M KCl to remove column-bound impurities. Bound proteins were eluted in a linear gradient of 28 CVs to elution buffer A, containing 250 mM imidazole using an ÄKTA purifier 10 (GE Healthcare, Freiburg, Germany). Protein containing fractions were pooled and either digested with 200 U thrombin (Sigma-Aldrich, Steinheim, Germany) at 277 K overnight (added 4 mM $CaCl_2$ to the buffer for the digest) to remove the N-terminal 6xHis-Tag or directly diluted in buffer A_{hep} and loaded on a 1 ml HiTrap heparin column (GE Healthcare, Freiburg, Germany), equilibrated with buffer A_{hep} using an ÄKTA purifier 10 at 277 K. In case of loading the thrombin-digested protein on the heparin column, a HiTrap Benzamidine column (GE Healthcare, Freiburg, Germany) was connected ahead, to remove thrombin during the heparin column loading process. Digested or undigested proteins were eluted in a linear gradient of 28 CVs to buffer B_{hep}. Protein containing fractions were diluted in buffer A_{monos} and loaded on a MonoS 5/50 GL column (GE Healthcare, Freiburg, Germany), equilibrated in buffer A_{monos}. Again bound proteins were eluted in a linear gradient of 28 CVs to buffer B_{monos} and protein fractions were concentrated using Amicon Ultra 10 KDa MWCO (10 K) filters (Millipore, Darmstadt, Germany). Further purification was achieved by gel-filtration chromatography on a Superdex S75 10/300 GL column (GE Healthcare, Freiburg, Germany), equilibrated in storage buffer. Fractions containing pure SPRY protein were pooled, concentrated in Amicon Ultra 10K filters and stored in ~ 20 mg/ml (700-800 µM) aliquots at 193 K for crystallization. All purification steps were verified by 15 % (w/v) SDS-PAGE, purified SPRY protein was confirmed to run as a single band with a electrophoretic mobility similar to a 28 kDa particle and identity was further controlled in MALDI-MS experiments conducted by Melanie Müller and Marion Gradl (**see section 2.3.7**).

Selenomethionine-substituted protein was expressed according to Van Duyne,1993[151] (**see section 2.1.5**) and protein purification was performed essentially as described for native protein, except that DTE concentration in the buffers A_{hep}, A_{monos} and storage buffer was increased from 3 mM to 5 mM.

2.3.5 Purification of HSPC117 complex components

Besides DDX1 three other protein components of the pentameric HSPC117 complex were expressed, the proteins Fam98b, CGI-99 and ASW.

Materials and Methods

The cleared supernatant of Fam98b or CGI-99 expressing bacteria lysate was loaded on 1 ml of Co^{2+}-Talon beads (Clontech, Saint-Germain-en-Laye, France), pre-equilibrated with lysis buffer A using a peristaltic pump P-1 (GE Healthcare, Freiburg, Germany) at 277 K. After loading, beads were washed with three CVs of lysis buffer A and proteins were eluted in a linear gradient of 28 CVs to lysis buffer A containing additional 120 mM Imidazole. Protein Fam98b could not be found in the eluate fractions and purification was ceased at that point; however, soluble protein material of CGI-99 was detected. CGI-99 protein containing fractions were diluted in buffer A_{monoQ} and loaded on a MonoQ 10/100 GL column (GE Healthcare, Freiburg, Germany), equilibrated with buffer A_{monoQ}. Bound proteins were eluted in a linear gradient of 28 CVs to buffer B_{monoQ} and CGI-99 protein containing fractions were concentrated using Amicon Ultra 10K centrifugal filter units (Milipore, Darmstadt, Germany) with a MWCO of 10 kDa. Finally, the proteins were polished by gel-filtration chromatography on a Superdex S75 10/300 GL column (GE Healthcare, Freiburg, Germany), equilibrated in storage buffer. Fractions containing pure CGI-99 protein were pooled, concentrated in Amicon Ultra 10K filters and stored in ~ 10 mg/ml (300-400 µM) aliquots at 193 K.

The cleared supernatant of ASW expressing bacteria lysate was loaded on a 1 ml Ni^{2+}-NTA HiTrap column (GE Healthcare, Freiburg, Germany) pre-equilibrated with lysis buffer A using a peristaltic pump P-1 (GE Healthcare, Freiburg, Germany) at 277 K. The column was washed with three CVs of lysis buffer A and with one CV of lysis buffer A containing additionally 1 M KCl to remove column-bound impurities. Bound proteins were eluted in a linear gradient of 28 CVs to elution buffer A, using an ÄKTA purifier 10 (GE Healthcare, Freiburg, Germany). ASW protein containing fractions were pooled, diluted in buffer A_{hep} and loaded on a HiTrap 1 ml heparin column (GE Healthcare, Freiburg, Germany), equilibrated with buffer A_{hep} using an ÄKTA purifier 10 at 277 K. Bound ASW protein was eluted in a linear gradient of 28 CVs to buffer B_{hep} and protein containing fractions were concentrated using Amicon Ultra 10 KDa MWCO filters (Millipore, Darmstadt, Germany). The concentrated protein sample was loaded on a Superdex S75 10/300 GL column (GE Healthcare, Freiburg, Germany), equilibrated in storage buffer. Fractions containing pure ASW protein were pooled, concentrated in Amicon Ultra 10K filters and stored in ~ 30 mg/ml (300-400 µM) aliquots at 193 K.

2.3.6 Co-purification and pulldown studies

To test for interactions between the individual subunits of the HSPC117 complex, co-expression and co-purification studies were performed. In co-expression experiments, cell pellets of DDX1, CGI-99 and Fam98b expressing bacteria (pellet from 2 l LB medium per construct) were combined and cell walls were disrupted by pulse-sonication (Branson Sonifier, Dietzenbach, Germany) at 277 K for 20 min. The combined lysate was cleared by ultracentrifugation at 125.000 x g in a L8-70M ultracentrifuge with a 45Ti rotor (Beckman Coulter, Krefeld, Germany) at 277 K for 40 min. The cleared supernatant, putatively containing the three recombinant proteins, was loaded on a 1 ml Ni^{2+}-NTA HiTrap column (GE Healthcare, Freiburg, Germany) pre-equilibrated with lysis buffer A using a peristaltic pump P-1 (GE Healthcare, Freiburg, Germany) at 277 K. The column was washed with three CVs of lysis buffer A and with one CV of lysis buffer A containing additionally 1 M KCl to remove column-bound impurities. Bound proteins were eluted in a linear gradient of 28 CVs to elution buffer A, using an ÄKTA purifier 10 (GE Healthcare, Freiburg, Germany). Proteins not forming a complex were further purified by individual protocols as described above.

A pulldown experiment was performed with the proteins ASW and DDX1.

The cleared supernatant of ASW expressing bacteria lysate was loaded on a 1 ml Ni^{2+}-NTA HiTrap column (GE Healthcare, Freiburg, Germany) pre-equilibrated with lysis buffer A using a peristaltic pump P-1 (GE Healthcare, Freiburg, Germany) at 277 K. The column was washed with three CVs of lysis buffer A and with one CV of lysis buffer A containing additionally 1 M KCl to remove column-bound impurities. Bound proteins were eluted in a linear gradient of 28 CVs to elution buffer A, using an ÄKTA purifier 10 (GE Healthcare, Freiburg, Germany). ASW protein containing fractions were pooled, diluted in ~ 10 ml of lysis buffer A (supplemented with 4 mM $CaCl_2$) and digested with 200 U thrombin (Sigma-Aldrich, Steinheim, Germany) at 277 K overnight to remove the N-terminal 6xHis-Tag. The thrombin-digested ASW protein was concentrated to ~ 2 ml in Amicon Ultra 10 KDa MWCO filters (Millipore, Darmstadt, Germany) and then diluted in 20 ml lysis buffer A, to decrease the concentration of perturbing imidazole in the sample. ASW protein sample was loaded via a 50 ml superloop™ (GE Healthcare, Freiburg, Germany) on a 1 ml Ni^{2+}-NTA HiTrap column (GE Healthcare, Freiburg, Germany) pre-equilibrated with lysis buffer A, with an upstream HiTrap Benzamidine column (GE Healthcare, Freiburg, Germany) using a peristaltic pump P-1 (GE Healthcare, Freiburg, Germany) at 277 K. The Ni^{2+}-NTA HiTrap column was washed with three column volumes (CVs) of lysis buffer A. The flow-through and the wash were collected that should contain only the 6xHis-Tag free ASW protein sample.

The cleared supernatant of DDX1 expressing bacteria lysate was loaded on a 1 ml Ni^{2+}-NTA HiTrap column (GE Healthcare, Freiburg, Germany) pre-equilibrated with lysis buffer A using a peristaltic pump P-1 (GE Healthcare, Freiburg, Germany) at 277 K. The column was washed with three CVs of lysis buffer A and then 6xHis-Tag free ASW protein sample was loaded on the same column. Bound proteins, putatively forming a complex, were eluted in a linear gradient of 28 CVs to elution buffer A using an ÄKTA purifier 10 (GE Healthcare, Freiburg, Germany).

Materials and Methods

2.3.7 Determination of protein concentration

The concentration of all protein samples was determined spectroscopically by measuring the absorbance at a wavelength of 280 nm in a NanoDrop ND-1000 spectrophotometer (NanoDrop Technologies, Wilmington, USA). At this wavelength, the aromatic amino acids tyrosine and tryptophan as well as cystein determine the overall absorbance[152, 153]. Based on amino acid composition individual extinction coefficients (ε) per construct were calculated using the online tool ProtParam[154]. Protein concentrations were determined according to the law of Lambert-Beer:

$$A_{280} = \lg\left(\frac{I_0}{I_1}\right) = \varepsilon_{280} \cdot C \cdot d \quad \rightarrow \quad C\left[\frac{mol}{l}\right] = \frac{A_{280}}{d \cdot \varepsilon_{280}} \qquad \text{[Equation 1]}$$

In this equation I_0 denotes the intensity of the incoming light, I_1 the intensity of the transmitted light, ε_{280} the respective extinction coefficient at 280 nm, A_{280} the protein absorbance at 280 nm and d denotes the pathlength of the cuvette in cm.

2.3.8 Mass spectrometry

All Mass spectrometric analysis was done by either Melanie Müller (M. M.) or Marion Gradl (M. G.). Sample handling and sample preparation for the measurements were performed according to Wulfmeyer and colleagues (2012)[155] by M.M. and M.G..

Electrospray-ionisation (ESI) measurements were performed by M. M. and M. G. using desalted samples. Samples were analyzed in a Bruker maXis hybrid quadrupole/atmospheric pressure ionization orthogonal accelerated time of flight (TOF) mass spectrometer, equipped with a reflectron and an Apollo II ESI source (Bruker Daltoniks, Billerica, USA). Data were collected with the standard Bruker micrOTOF control version 3.0 SR1 compass 1.3 SR2 software and analyzed with the data analysis version 4.0 SP3 ESI compass 1.3 software. Instrument calibration was conducted with the appropriate components (m/z 118-1222) of the ESI-L standard (Agilent Technologies, Waldbronn, Germany) and optimized for m/z range of ca. 100-1000. Concentrations of the samples were adjusted to a final concentration of 10-60 µM in 50 % acetonitrile, 49.9 % water and 0.1 % formic acid. Using a syringe pump (KD Scientific, Holliston, USA) samples were injected into the mass spectrometer at a flow rate of 3 µl/min. Samples were analyzed in both positive and negative polarity in MS and MS/MS modes. MS/MS was performed with low energy collision-induced dissociation (CID), using various collision energies ranging from 5 to 50 V.

Peptide mass fingerprinting (PMF) was performed by M. M. and M. G. by trypsin digestion of proteins "in-gel" and then applying standard procedures on the obtained peptides[155]. Matrix-assisted laser desorption/ionisation (MALDI)-TOF with a Shimadzu Axima Performance mass spectrometer

Materials and Methods

(Shimadzu/Kratos Analytical, Duisburg, Germany) in reflectron mode was used to analyze the peptides and the same instrument was used to confirm identities of select peptides by sequencing with high energy CID MS/MS. The Mascot PMF or Mascot MS/MS search software (Matrix Science, London, UK) with a hybrid taxonomy which includes all *E. coli* proteins and the specific proteins of interest in the BMM department was used to identify the proteins. Also see Mutschler 2011[156].

2.3.9 Domain mapping by limited proteolysis experiments

Limited proteolysis experiments were conducted by digesting protein samples with commercially available proteases (Sigma-Aldrich, Taufkirchen, Germany). 25 µl protein sample in storage buffer with a concentration of 70 µM was supplemented with 4 mM $CaCl_2$ and either 2.5 µg trypsin, chymotrypsin or thermolysin were added (0.1 g/l final protease concentration). Digests were incubated at 310 K and quenched at defined time-points by freezing 2 µl aliquots in liquid nitrogen. Digestion products were separated via SDS-PAGE and defined proteins bands were excised from the gel and analyzed by MALDI-MS.

2.3.10 Electrophoretic mobility shift assay

Electrophoretic mobility shift assays (EMSAs) were used to test binding of DDX1 to different RNA species. Binding to a 10mer polyA, a 10mer polyU and a 13mer RNA of random sequence was tested by using 3'-Carboxyfluorescein (FAM) labeled variants of the RNAs. RNA species with the FAM label at the 5'-end were also tested, but did not differ in binding when compared with the 3'-labeled variant, so that experiments were restricted to the following RNA species:

10mer polyA RNA	5'-AAAAAAAAAA-3'-FAM
10mer polyU RNA	5'-UUUUUUUUUU-3'-FAM
13mer RNA of random sequence (according to Jankowsky and Fairman, 2008)[157]	5'-AGCACCGUAAAGA-3'-FAM
20mer RNA	5'-GCGAGACAGUGUGACUUUGG-3'-FAM

all RNAs were purchased from biomers.net GmbH (Ulm)

For the binding reactions, 0.5 µM RNA was incubated with 0.5 to 50 µM protein at 293 K in 20 µl EMSA reaction buffer (see section 2.1.2) for 20 min. Reactions were quenched by cooling them to 277 K and 8 % (v/v) of glycerol were added. Reactions were separated by native PAGE on a 5 %/15 % step-gradient acrylamide-gel (see section 2.3.2). Shifted RNA was visualized by excitation of the the FAM label at 302 nm on a UV-scanner. Subsequently, proteins were stained using InstantBlue.

2.3.11 Helicase assay

A gel-based helicase assay was used to test for an unwinding activity of DDX1. For this assay double-stranded substrate was prepared by annealing either short, fluorescently labeled DNA oligonucleotides with RNA or labeled RNA with RNA oligonucleotides. All substrate strands – unlabeled or fluorescently labeled with either fluorescein (Fl) or carboxyfluorescein (FAM) – were ordered as chemically synthesized oligonucleotides from biomers.net (Ulm, Germany).

DNA – 41mer (according to Li et al., 2008)[111]	5'-GAATACAAGCTTGCATGCCTGCAGGTCGACTCTAGAGGATC-3'-Fl
RNA – 29mer (according to Li et al., 2008)[94]	5'-GAUCCUCUAGAGUCGACCUGCAGGCAUGC-3'
DNA – 13mer (according to Jankowsky and Fairman, 2008)[157]	5'-AGC ACC GTA AAG A-3'-Fl
RNA – 38mer (according to Jankowsky and Fairman, 2008)[157]	5'- UCUUUACGGUGCUUAAAACAAAACAAAACAAAACAAAA-3'
RNA – 13mer (according to Jankowsky and Fairman, 2008)[157]	5'-AGCACCGUAAAGA-3'-FAM
all oligonucleotides were purchased from biomers.net GmbH (Ulm)	oligos were labeled either with Fluorescein (=Fl) or Carboxyfluorescein (=FAM) at their termini

Annealing reactions were conducted in 25 µl annealing buffer (see section 2.1.2) with 100 µM unlabeled and 50 µm labeled oligonucleotide, to ensure that all labeled RNA is incorporated in a duplex. Annealing reactions were heated to 368 K for 2 min and then incubated at RT for 2 h. For the actual unwinding experiment 1.25 µM duplex substrate was incubated with 10 – 20 µM DDX1 and 10 mM $MgCl_2$ (which was replaced by 10 mM EDTA in control reactions) in a total volume of 20 µl unwinding buffer (see section 2.1.2). Optional 10 mM of ATP, AppNHp, ATPγS or ADP (Jena Bioscience, Jena, Germany) were added. Reactions were incubated at 310 K for 30 min and then quenched by cooling to 277 K and addition of 5 µl RNA loading dye. Unwound substrate strands were then separated by native PAGE on a 15 % (w/v) native gel prepared in 1xTBE buffer pH 8.3 (see section 2.3.2). Separated fluorescently labeled oligonucleotide strands were visualized by excitation of the Fl or FAM label at 302 nm using a UV-scanner. Subsequently, proteins were stained using InstantBlue.

2.4 Biophysical Methods

2.4.1 Circular dichroism spectroscopy

Circular dichroism (CD) spectroscopy was used to estimate the secondary structure composition of protein constructs and their thermal stability. All measurements were performed with 5 µM protein sample in storage buffer (see section 2.1.2) in a Jasco J-810 CD spectropolarimeter (Jasco Corp., Groß-Umstadt, Germany). Spectra of the protein samples were recorded from 180 nm to 260 nm in a quartz-glass cuvette (0.1 cm path length, Hellma, Müllheim, Germany) in a total volume of 200 µl storage buffer. A buffer baseline was recorded for all measurements, to correct for absorption of HEPES in the storage buffer (see section 2.1.2). The signal output from the instrument in [mdegs] was converted into mean residue ellipticity (θ_{MRE}), by first calculating the mean residue weight (MRW):

$$MRW = \frac{molecular\ weight}{[no.of\ residues]-1} \qquad \text{[Equation 2]}$$

and then calculating the θ_{MRE} according to

$$\theta_{MRE} = \frac{[m \deg s] \cdot MRW}{d \cdot C} \quad \left[\frac{\deg \cdot cm^2}{dmol}\right] \qquad \text{[Equation 3]}$$

with d = path length of the cuvette in [mm] and C = protein concentration in [mg/ml]

For thermal denaturation curves, the cuvette was heated from 293 K to 368 K with a heating rate of $\Delta T/\Delta t$ = 1 K/min and unfolding of protein secondary structure was followed by recording light-polarisation at 222 nm. Melting curves were fitted to a two-state-unfolding equation according to Consalvi and colleagues[158, 159]:

$$signal = \frac{aN + bN \cdot T + (aD + bD \cdot T) \cdot \exp[\Delta H/R \cdot (1/T_m - 1/T)]}{1 + \exp[\Delta H/R \cdot (1/T_m - 1/T)]} \qquad \text{[Equation 4]}$$

where aN and aD are the baseline intercepts of the native (N) and denatured (D) state, bN and bD the corresponding slopes of the baselines, ΔH the enthalpy at the transition midpoint, T the current temperature, T_m the melting temperature and R the gas constant.

2.4.2 Static light scattering

The oligomeric state of the protein was assessed via static light scattering (SLS). For static light scattering 50 µM of DDX1 protein in 40 µl storage buffer was injected on a Superdex 200 10/300 GL column (Pharmacia, Freiburg, Germany) using a Waters FPLC system (Waters Corp., Milford, USA). The Superdex 200 column was connected to a Wyatt Dawn Heleos II multi-angle-light-scattering (MALS) detector (Wyatt Technology, Dernbach, Germany), detecting scattered light from 18 angles. Refractive index and light scattering data were used to calculate the radius of gyration and the corresponding molecular weight with the help of the Astra Software (Wyatt Technology, Dernbach, Germany).

2.4.3 Dynamic light scattering

For dynamic light scattering (DLS) 25 µM of DDX1 protein in a total volume of 60 µL storage buffer were measured in a Viscotek 802 (Malvern Instruments, Mannheim, Germany), which records scattered light at 90° angle. Before each measurement the sample was centrifuged at maximal speed in a table-top-centrifuge (Heraeus® Biofuge®, Thermo-Scientific, Waltham, MA, USA). 30 light fluctuation curves with four second measurement time each were recorded. All traces with constant intensity were averaged to fit a combined auto-correlation function. From the auto-correlation function the hydrodynamic radius was extracted and used to calculate the molecular weight assuming a globular protein shape. All DLS data analysis was performed using the OmniSIZE Software (Malvern Instruments, Mannheim, Germany).

2.4.4 Steady-state ATPase assay

Steady-state ATP hydrolysis was measured using a coupled photometric assay[160, 161]. The rate of hydrolysis of ATP by DDX1 is measured directly by a conversion cascade that uses the produced ADP. In this coupled assay ADP is converted to ATP by pyruvate-kinase (PK), thereby converting phosphoenolpyruvate (PEP) to pyruvate (PYR). Produced PYR is further utilized by lactate-dehydrogenase (LDH) which converts it to lactate (LAC) and oxidizes NADH to NAD^+. Thereby, the decrease of NADH absorbance at 340 nm can be used to directly quantify the molar amount of produced ADP. Reactions (100 µl) were measured at 298 K in a Jasco V-650 spectrophotometer (Jasco Corp., Groß-Umstadt, Germany).

For the reactions 2 x ATPase buffer (see section 2.1.2) was supplemented with a mixture of 4 U/ml LDH / 2.8 U/ml PK (Roche, Mannheim, Germany), 15 mM KCl and 1 mM ATP. Reaction was started by addition of 5 mM $MgCl_2$ and was incubated until a stable baseline was reached, indicative of a complete conversion of traces of ADP contaminations into ATP. For the start of the reaction protein and substrate were added and rates of ADP generation (assumed to equal the ATP hydrolysis rate) were measured.

The initial reaction rates were determined as

$$v = \frac{\frac{\Delta Abs_{340}}{\Delta t}}{\varepsilon_{NADH} \cdot d \cdot c}$$ [Equation 5]

with v = intial reaction rate, ε_{NADH} = extinction coefficient of NADH at 340 nm (6220 M^{-1} cm^{-1}), d = path length of cuvette, c = protein concentration, ΔAbs_{340} = decrease in absorbance at 340 nm, Δt = time interval.

The rate for $\Delta Abs_{340}/\Delta t$ was determined from a linear fit to the steady state decrease in absorbance using the Spectra-Viewer software (Jasco Corp., Groß-Umstadt, Germany).

To determine the Michaelis-Menten parameters, 1 µM DDX1 was incubated with varying amounts of either ATP-MgCl$_2$ or a 20mer RNA (5'-GCGAGACAGUGUGACUUUGG-3') in ATPase buffer. The ATP-concentration dependence of the ATPase rate and stimulation of hydrolysis by RNA titration were analyzed with the hyperbolic form of the Michaelis-Menten-equation.

$$v = v_{max} \cdot \frac{S}{S + K_m}$$ [Equation 6]

with S = substrate concentration, v_{max} = maximal apparent reaction velocity, K_m = apparent Michaelis-Menten constant.

Measurements with the Walker A lysine-mutant of DDX1 (K52A) were performed at varying ATP-MgCl$_2$ and protein concentrations in the presence or absence of stimulating RNA.

2.4.5 Fluorescence equilibrium titrations

Binding affinities for nucleotides were determined by fluorescence equilibrium titration, using N-methylanthraniloyl(mant)-labeled nucleotide analogues. Mant-ADP and mant-dADP were purchased from BIOLOG (Bremen, Germany). Fluorescence spectra of isolated and protein bound mant-nucleotides in titration buffer (see section 2.1.2) were recorded on a Jasco FP-8500 spectrofluorometer (Jasco Corp., Groß-Umstadt, Germany) thermostated to 298 K using a peltier-element. 50 nM up to 500 nM mant-nucleotide in 150 µl titration buffer (depending on measurement) were titrated with DDX1 in a range of 10 nM to 5 µM. For competition experiments with unlabeled nucleotides, the concentration of mant-nucleotide (200 nM or 500 nM) and protein (1 µM) were kept constant, and 0.1 µM to 120 µM ADP-MgCl$_2$ or 5 µM to 5 mM ATP- MgCl$_2$ were titrated.

Fluorophores were excited at 356 nm and spectra were recorded from 370 nm to 520 nm with 5 nm slid-width of excitation and emission light. The fluorescence of equilibrated solutions at 488 nm emission in

Materials and Methods

dependence of protein concentration was used to plot binding curves. Titration data of protein binding to mant-nucleotides were fitted to a quadratic binding equation[162, 163]:

$$signal = F_0 + F_{max} \cdot \frac{(X + B_0 + K_d)/2 - \sqrt{[(X + B_0 + K_d)/2]^2 - X \cdot B_0}}{B_0}$$ [Equation 8]

where F_0 is the baseline, F_{max} the amplitude of signal change, B_0 is the total concentration of mant-nucleotide that was constant during the measurement (B_0 = B+AB), X is the concentration of component A (=DDX1-protein) that is varied and K_d is the dissociation constant for the complex AB of mant-nucleotide and DDX1.

Data, recorded during competition experiments, where mant-nucleotide-DDX1 complex was incubated with excess of unlabeled nucleotides were fitted to a cubic binding equation[164, 165] with [AC] as the concentration of the complex of protein with unlabeled nucleotide :

$[AC]^3 + a_1[AC]^2 + a_2[AC] + a_3 = 0$, with the coefficients

$a_0 = K_{d,AB} - K_{d,AC}$

$a_1 = ([A_0](K_{d,AC} - K_{d,AB}) + [B_0](2K_{d,AC} - K_{d,AB}) + [C_0]K_{d,AB} - K_{d,AB}^2 + K_{d,AB}K_{d,AC})/a_0$

$a_2 = ([A_0][B_0](K_{d,AB} - 2K_{d,AC}) - [B_0]^2 K_{d,AC} - [B_0]K_{d,AB}([C_0] + K_{d,AC}))/a_0$

$a_3 = \frac{[A_0][B_0]^2 K_{d,AC}}{a_0}$ [Equation 9]

where [C_0] is the total concentration of unlabeled nucleotide, [B_0] the total concentration of mant-nucleotide and [A_0] the total concentration of protein. The general solution of a cubic equation (see Press et al. 1989[166]) gives three solutions, out of which only one is meaningful[164] and was selected manually.

The $K_{d,AB}$ for the protein mant-nucleotide complex was used as a constant as previously determined from the binding titrations. Data fitting was performed either with GraFit 5.0 (Erithacus Software, Horley, UK) or Prism 5.0 (GraphPad Software, San Diego, USA). All titrations involving ATP were performed in the presence of 2 mM PEP and 4U/ml PK (Roche, Mannheim, Germany) as an ATP regeneration system. The influence of RNA on binding parameters was tested by conducting the titrations in the presence of 28 µM RNA. Note that the protein-stock used for titrations was incubated and saturated with excess amounts of RNA (2 mM) prior to measurements. Three different RNAs were tested for their influence on binding parameters, a 10mer polyA RNA, a 13mer RNA of random sequence and a 20mer RNA (see table in section 2.3.8 - EMSA). All RNAs were purchased from biomers.net GmbH (Ulm, Germany).

2.4.6 Stopped-flow measurements

Transient kinetic measurements were used to check the reliability of the equilibrium titrations. Data were recorded on a BioLogic SFM-400 stopped-flow instrument (BioLogic Science instruments, Grenoble, France), thermostated to 298 K. Mant-fluorescence was excited at 356 nm and detected with a 420 nm cut-off filter (420FG03-25, LOT Oriel Group, Darmstadt, Germany). Traces were recorded in triplicates and averaged. For binding experiments 0.5 µM DDX1 in titration buffer was rapidly mixed with 2.5 to 7.5 µM mant-nucleotides in a measurement chamber, consisting of a FC-15 (1.5 x 1.5 mm) quartz cuvette (BioLogic, Claix, France) and traces were fitted to the sum of two exponential functions. The first rate constant that made for the majority of the amplitude did scale linearly with nucleotide concentration. The on-rates for nucleotide binding were obtained from the slope of a plot of those first rate constants. Off-rates for nucleotide binding were obtained by chase experiments. Therefore, 0.5 µM mant-nucleotide were pre-incubated with 0.5 µM DDX1 in titration buffer (see section 2.1.2) and then displaced by rapid mixing with either 250 or 500 µM ADP. Traces were fitted to a single exponential function that did contain the rate constant for the mant-nucleotide off-rate. The dissociation constant K_d was calculated from the ratio of off- to on-rate. Experiments were performed with both mant-ADP and mant-dADP.

Recorded stopped-flow traces were initially analysed analytically assuming pseudo-first order conditions. For the titration of the proteins with fluorescent nucleotides, the nucleotide was assumed to be in large excess and therefore its concentration unaltered during the binding reaction.

$$P + mant \underset{k_{-1}}{\overset{k_1}{\rightleftarrows}} Pmant$$

$$\frac{dPmant}{dt} = k_1 \cdot [mant] \cdot [P] - k_{-1} \cdot [Pmant] = k'_1 \cdot [P] - k_{-1} \cdot [Pmant] \qquad \text{[Equation 10]}$$

This differential equation can be solved by an exponential function:

$$[Pmant] = Amp \cdot e^{-(k'_1 + k_{-1}) \cdot t} \qquad \text{[Equation 11]}$$

In order to obtain positive amplitudes data traces were fitted to the sum of exponential functions in the form

$$signal = Amp \cdot (1 - e^{-k \cdot t}) + offset \qquad \text{[Equation 12]}$$

Fitting of the data traces to exponential functions was done with GraphPad Prism 5.01 (GraphPad Software, San Diego, USA).

2.4.7 RNA affinity accessed by equilibrium titrations

RNA affinities of DDX1 were determined by exploitation of the cooperativity in ATP and RNA binding and using mant-nucleotide equilibrium titrations as described before (see section 2.4.5). Mant-dADP

(0.2 µM) was incubated with DDX1 (1 µM) in titration buffer for complex-formation and subsequently this complex was partially displaced upon addition of either 50, 100, 200 or 400 µM ATP-MgCl$_2$ in individual experiments. In the actual experiment, RNA was titrated to the complex and binding was followed by fluorescence decrease, induced by RNA dependent ATP binding. Again an ATP regeneration system was included in all measurements. The RNA-dependent mant-displacement curves at different ATP concentrations were fitted globally by numeric iteration using DynaFit (Petr Kuzmic), using a minimal binding scheme as reaction mechanism for modeling (see appendix 7.6 for DynaFit script file). All previously determined equilibrium titration data were included in the fit.

2.5 Crystallographic methods

2.5.1 Protein crystallization

Crystallization experiments were performed as sitting-drop vapour diffusion setup using commercially available crystallization-screens (Qiagen, Hilden, Germany) at 293 K. Experiments were set-up in 96-well XTL low profile plates (Greiner Bio-One, Frickenhausen, Germany) by mixing 100 nl protein sample at 20 mg/ml with 100 nl reservoir solution using a mosquito robotic system (TTP Labtech, Hertfordshire, UK). X-ray diffracting crystals were obtained with the constructs of the SPRY domain (see **table 2.2**).

Table 2.2 SPRY domain crystal-forming constructs
Details on the crystallization conditions are given; a "+"-symbol in the last column, means linbro-optimization was performed

construct	final concentration in crystallization experiment	condition	time of crystal growth	Linbro-optimization
SPRY_72-283	21 mg/ml	40 % (v/v) PEG 600, 0.1 M tri-Sodium-citrate pH 5.5	3 days	+
SPRY_72-283ΔTag	25 mg/ml	35% (v/v) PEG 600, 0.1 M tri-Sodium-citrate pH 5.5	6 days	+
SPRY_84-261	19 mg/ml	30% (w/v) PEG 3000, 0.1 M CHES pH 9.5	2 days	-
SPRY_84-261 (SeMet)	24 mg/ml	30% (w/v) PEG 3000, 0.1 M CHES pH 9.5	2 days	+

Initial crystal-formation with the SPRY_72-283 protein was observed after three days in a reservoir solution consisting of 40 % (v/v) polyethylene glycol (PEG) 600, 0.1 M tri-Sodium-citrate pH 5.5. This condition was refined by hanging-drop vapour diffusion using 24-well Linbro plates (Greiner Bio-One, Frickenhausen, Germany), mixing 1 µl protein solution with 1 µl of reservoir solution. Small crystals, which were obtained after three days, were crushed and used to streak-seed into fresh protein/reservoir drops.

Crystals of protein SPRY_84-216 were obtained in various conditions using the JCSG Core Suite (Qiagen, Hilden, Germany). Largest and best diffracting crystals grew in 30% (w/v) PEG 3000, 0.1 M CHES pH 9.5. Crystals of protein SPRY_84-216 with Selenomethionine incorporated were obtained under the same conditions.

Protein SPRY_72-283 lacking the N-terminal hexahistidine Tag (SPRY_72-283ΔTag) showed crystal formation in a 96-well screen with a reservoir solution containing 35% (v/v) PEG 600, 0.1 M tri-Sodium-citrate pH 5.5. The 6xHis-Tag had been cleaved-off proteolytically by thrombin, removing the sequence MGSSHHHHHHSSGLVPR from the N-terminus. Spheroid crystals appeared after three days and grew as single crystals with typical dimensions of 140 x 90 x 40 µm within six days. For the ease of crystal handling, the experiment was reproduced in a hanging-drop vapour diffusion set-up using 24-well Linbro plates (Greiner Bio-One, Frickenhausen, Germany). In each well 1 µl of protein solution was overlaid with 1 µl of reservoir solution on a coverslide and incubated hanging, equilibrated against a total volume of 700 µl reservoir/precipitant solution. The same buffer composition as in the original hit condition in the 96-well plate was used and Linbro plates were incubated at 297 K for three days.

2.5.2 Crystal harvesting and cryoprotection

For data collection, single crystals were harvested into a cryoloop. To harvest the SPRY crystals from Linbro plates 0.2 mm loops, equipped with a 20 µm nylon thread were used. For cryoprotection of crystals from protein constructs SPRY_72-283 and SPRY_72-283ΔTag the mother solution was used directly, whereas for the SPRY_84-216 protein 20 % (v/v) glycerol were added to the mother solution. Crystals were transferred with the cryoloop two times into 3 µl of cryoprotectant solution and incubated briefly during each step. Subsequently they were flash-cooled in liquid nitrogen.

2.5.3 Data collection and processing

Diffraction data were collected at synchrotron beamline X10SA (PXII) at the Swiss Light Source (SLS) of the Paul-Scherrer-Institute in Villigen, Switzerland. Collection was done in a joint venture between Scientist from Heidelberg and Dortmund, which formed a "data-collection" team. Data were collected at 100 K in a rotation setup on a Pilatus 6M detector (Dectris, Baden, Switzerland).

Prior to synchrotron measurements, crystals were tested for their diffraction quality in-house. A Rigaku MicroMax 007 HF (Rigaku, Kemsing, UK) microfocus rotating anode (40 kV/30 mA), equipped with

VariMax HF focussing mirrors (Confocal Max-Flux® CMF optics, Rigaku, Kemsing, UK) and a Mar345 imaging plate (MarResearch, Norderstedt, Germany) were used. Cooling to 100 K was provided by a nitrogen cryostream (Oxford Cryosystems, Long Hanborough, UK).

All data sets were processed with the XDS software package[167, Kabsch, 2010 #244]. Diffraction data were indexed and integrated with XDS and then scaled and merged using XSCALE[167]. Finally, XDSCONVERT was used to convert files into the appropriate format compatible with the programs of the CCP4 suite[168], for phasing, model building and refinement. To test for model bias, 5 % of the reflections were randomly picked and omitted during refinement and used for calculation of an R_{free} value.

R_{meas}, calculated by XDS was used as an indicator of data consistency. R_{meas} corresponds to R_{merge}, corrected by the redundancy, according to [169]

$$R_{meas} = \frac{\sum_{hkl} \sqrt{n/(n-1)} \sum_{i} |\langle I_{hkl} \rangle - I_{hkl,i}|}{\sum_{hkl} \sum_{i} I_{hkl,i}}$$ [Equation 13]

with n as the redundancy/multiplicity, <I_{hkl}> as the mean intensity of symmetry-equivalent reflections and i as a numerator (i.e. giving the sum over all symmetry-equivalent reflections).

Resolution bins, where the R_{meas} [equation 13] were better than 50 % and the <I/δ(I)> was better than 7 were considered to be reliable.

2.5.4 Phasing by molecular replacement

Molecular replacement (MR) was used of to obtain initial phases for the SPRY crystals of the protein construct SPRY_72-283ΔTag. MR was performed using PHASER[170]. The SPRY domain of Ash2L (PDB entry 3TOJ[107]), deposited in the protein data bank (PDB), with a sequence identity of 23.8 % was used as a search model, but alternative conformations and water molecules were removed. The search was carried out using a resolution cut-off at 2.5 Å. With the determined unit cell of the crystal symmetry as obtained by XDS[171], the Matthews coefficient[172] of the crystal was calculated, according to

$$\frac{volume\ of\ unit\ cell}{molecular\ weight \cdot X \cdot Z}$$ [Equation 14]

where Z is the number of AUs in the unit cell and X is the number of molecules in the AU.

The number of molecules per AU was estimated from the Matthews coefficient, by comparison of X values with empirical observed ones[173] and used to constrain the MR search. Based on the Z-score of the rotation- and translation functions, possible solutions were evaluated.

2.5.5 Structure determination and refinement

Initial phases from MR were refined using the CCP4 interface[168, 174], by a rigid-body refinement using REFMAC[175]. A model of the SPRY domain was built manually using Coot [46]. This model was improved by iterative cycles of refinement with REFMAC using TLS[176] and manual improvement in Coot, leading to the final model. Manual model building in Coot was guided by secondary structure prediction[177] and residues that belong to secondary structure elements were placed first, due to easier allocation. Residues that were initially hard to place were first build as alanine residues.

To guide refinement, the R-factor (R_{work})[178] as calculated by REFMAC, was used as an indicator for the model quality.

$$R_{work, free} = \frac{\sum_{hkl} ||F_o| - |F_c||}{\sum_{hkl} |F_o|}$$ [Equation 15]

Model building was aimed towards minimizing the R_{work} to obtain a good agreement between the model and the measured data. Waters were picked manually by selecting peaks above 3 σ in the F_o-F_c difference map in correct hydrogen bonding distance. To differentiate between Na$^+$ and water the coordination sphere was considered (water – tetrahedral, Na$^+$ – octahedral). The R_{free} value, calculated during refinement from the excluded reflections, was reported[179]. The R_{free} value was calculated as described in equation 15. To antagonize model bias, the gap between R_{work} and R_{free} was minimized. The Model was considered to be sufficiently build at the point, when convergence between R_{work} and R_{free} i.e. a stable R-factor was reached.

The stereochemistry of the model for intermediate and final structures was evaluated using the subroutines implemented in Coot (rotamers, Ramachandran-plot...)[46]. Furthermore, MOLPROBITY[180] and PROCHECK[181] were used to assess the model quality. All structural representations were generated using PyMOL[182] with subsequent raytracing. Electrostatic surface potentials were calculated with the PyMOL plug-in APBS[183].

2.5.6 Homology modelling

The structural model of the entire human DDX1 helicase was assembled based on homology modeling of the helicase core using SWISS-MODEL[184]. The structure of a DEAD box RNA helicase from the hyperthermophile Methanococcus jannaschii (PDB entry 1HV8)[33] that shares 21.6 % sequence similarity to DDX1 was used as a template to model RecA-like domains 1 and 2. Both protein sequences were aligned and the alignment in FASTA-formatting was used as input parameter. The algorithm of *SWISS-MODEL*[184] initially mutates each residue in the template structure to the corresponding residue in the target protein to be modeled. Subsequently the geometry is optimized by energy minimization. In this way a homology model of the DDX1 helicase core without the SPRY domain was obtained.

This homology model was superposed on a structure of human DDX3X (PDB entry 2I4I)[185]. The SPRY domain was placed in the homology model according to the position of residues 250-260 of DDX3X with the help of *Coot*[46]. Finally the model[152, 160] was illustrated with PyMOL[182].

3. Results

3.1 Generation of a stable, recombinant protein constructs

This chapter describes the purification of recombinant protein material of *Homo sapiens* DDX1, suitable for structural, biophysical and biochemical studies. The sequence, coding for full-length protein was cloned from cDNA and this protein could be expressed in *E. coli* cells, but upon cell lysis several protein fragments were detected. Extensive construct optimization guided by biochemical and bioinformatic analysis led to the design of C-terminal truncated DDX1 variants that were stable and could be purified to high homogeneity. In addition, individual domains of DDX1 were cloned and purified. Constructs of the SPRY domain were prepared for structural studies. Several different SPRY constructs were generated to facilitate crystallization experiments later on. Moreover, further components of the HSPC117 complex (besides DDX1) were cloned and purified.

3.1.1 Characterization of DDX1 by bioinformatic tools

Initially, the DDX1 protein was analyzed *in silico* by bioinformatic tools to guide experimental design. Human DDX1[120] has a three domain architecture (**introduction Figure 1.10**) consisting of 740 amino acids with a molecular weight of 82.4 kDa according to the verified UniProt reference sequence (ID Q92499). A BLAST search revealed general conservation of DDX1 amongst mammalian orthologs[186]. No homolog of DDX1 was found in *Saccharomyces cerevisiae* and the closest relative in yeast, the helicase Dbp2p, shows only 18.7 % sequence identity (**suppl. Figure 7.1.2**, SIAS server, Pedro Reche[187]). The amino acid sequences of DDX1 orthologs of representative eukaryotic model organisms were analyzed in a multiple sequence alignment by Clustal Omega and a high conservation was found(**Figure 3.1.1**)[188]. The sequence identity between *H. sapiens* and *Caenorhabditis elegans*, the most distal homolog identified among these organisms, is 46.7 %. All helicase signature motifs are highly conserved between the model organisms (**Figure 3.1.1**), but in *C. elegans* the characteristic DEAD sequence of motif II[10] was found to be DEID. However, this third residue of motif II has been reported not to be strictly required for ATP hydrolysis in other DEAD-box proteins[15, 16]. Nevertheless, for the human export factor UAP56 it was shown, that a polymorphism of this residue can influence the stabilization of motif III (SAT)[26]. In the DDX1 polypeptide chain, the first conserved helicase motif Q is preceded only by few amino acids at the N-terminus. In contrast, at the C-terminus a long stretch of ~ 130 amino acids follows the last conserved helicase motif VI[92]. This suggests that DDX1 contains a C-terminal extension to the helicase core, which is further supported by comparison with other DEAD-box proteins[189], which revealed that the canonical helicase core ends at ~ residue 670.

Results

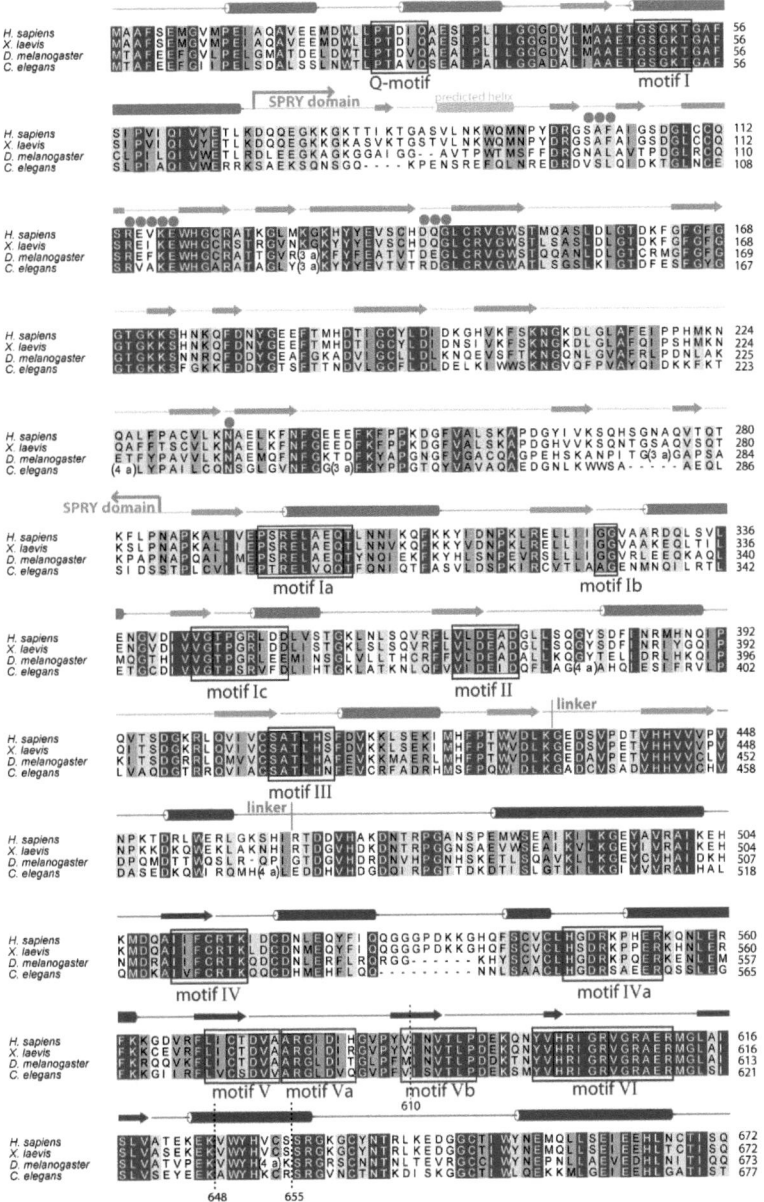

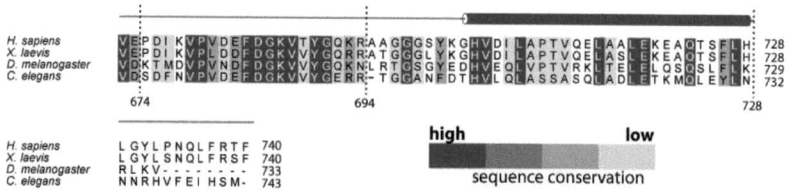

Figure 3.1.1 **Sequence alignment of DDX1 orthologs from eukaryotic model organisms.** DDX1 from *H. sapiens* is aligned with DDX1 orthologs from eukaryotic model organisms (no direct DDX1 homolog was found in yeast; *X. laevis*=*Xenopus laevis*) and the degree of sequence conservation is indicated by color coding. Secondary structure as predicted for human DDX1 with PSIPRED[177] is indicated above the sequence and colored according to domain allocation. For the SPRY domain information on the location of β-strands is inferred from the crystal structure. RecA-like domain 1 is colored in green, the SPRY domain is colored in blue and RecA-like domain 2 is colored in red. Conserved helicase sequence motifs (as explained in figure 1.3) are marked by blue bordering. Domain boundaries of C-terminally truncated constructs used in DDX1 expression trials are indicated by dashed lines with residue numbering.

The results of a secondary structure prediction of human DDX1 with PSIPRED[177] revealed that residues that are predicted to form helices and strands agree well with the fold of homologous DEAD-box proteins (also see **suppl. Figure 7.1.1**). The sequences of the predicted secondary structure elements, especially β-strands and α-helices, are mostly conserved across species, and only some loop regions show sequence variation. Additional sequence variation is found at the N- and C-terminus of the SPRY domain (see section on SPRY structure for details). Small differences in polypeptide chain length are detected in the *C. elegans* protein when compared to the human ortholog. Regions in the SPRY domain, in the loop proceeding motif II and in the linker between both RecA-like domains are longer in *C. elegans* DDX1 (**Figure 3.1.1**). Conversely the loop between motif IV and motif IVa is considerably shorter in *C. elegans* and *Drosophila melanogaster* DDX1 orthologs. Moreover, in *D. melanogaster* DDX1 the C-terminal tail that shows low sequence conservation is shorter by eight residues when compared to the human protein.

3.1.2 Expression and purification of recombinant DDX1

The coding sequence for human DDX1 was cloned from cDNA into a pET28a bacterial expression vector that encodes for an N-terminal 6xHis-Tag (**see materials and methods, section 2.2.9**). Upon expression in *E. coli*, the protein was found in the soluble fraction of the cell lysate. It was purified via Ni^{2+}-NTA affinity chromatography followed by another affinity chromatography step on a heparin column. Four prominent protein bands were observed on a SDS-gel of the heparin elution fractions (**Figure 3.1.2**). Bands were excised from the gel and their identity was determined via MALDI-MS PMF. The proteins, migrating in a band at around ~ 80 kDa did correspond to full-length DDX1 (calculated mass 82.4 kDa), the other bands, however, proofed to be DDX1 fragments (**Figure 3.1.2**). The presence of protein fragments could have different causes, for instance it could result from translation abortion or

Results

full-length protein might be proteolytically unstable and rapidly (partially) degraded by endogenous proteases of E.coli.

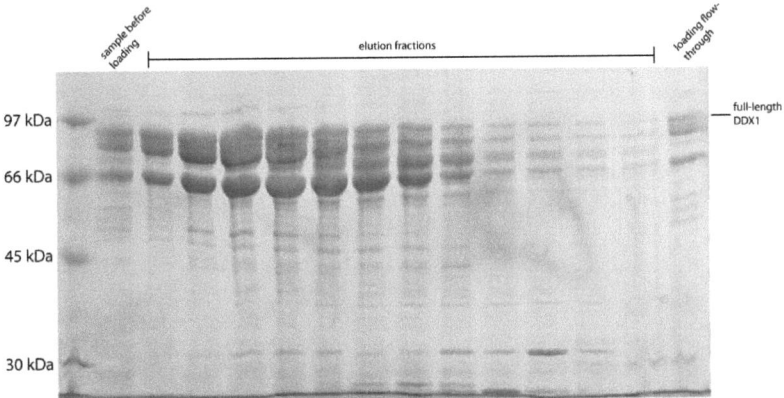

Figure 3.1.2 **DDX1 WT is soluble upon bacterial expression, but shows degradation.** Full-length DDX1 (amino acids 1-740) from *H. sapiens* was purified by Ni^{2+}- and heparin-affinity chromatography. Elution fractions from a heparin column run were separated on a 15% (w/v) SDS-gel and stained with Coomassie InstantBlue®. A band corresponding to full-length DDX1 is seen at ~80 kDa, but three additional bands are seen below, which may be indicative for abortive translation or degradation of the protein. Indeed, species running below the DDX1 full-length band did correspond to DDX1 fragments. Identity of all bands was checked via MALDI-MS.

Different strategies were used to prevent potential protein degradation. Protease inhibitor was added to the lysis buffer and the 6xHis-Tag was moved from the N- to the C-terminus. The protein was expressed in the periplasm using pET22b shuttling vector[135] and it was also expressed in different protease deficient *E. coli* strains[190] (see section 2.1.6). However, these attempts were not successful and the protein material was unsuitable for further *in vitro* studies of DDX1.

Analysis of the MALDI-MS PMF results showed that the observed protein fragments were prevalently truncated at the C-terminus. These fragments remained stable throughout the purification, which indicates that C-terminally truncated variants of human DDX1 may be less prone to proteolytic degradation when compared to the full-length protein. Based on this finding and guided by secondary structure prediction several C-terminally truncated protein constructs were designed (small excerpt in **Table 3.1**).

Table 3.1 **Truncated DDX1 constructs**
Different C-terminally truncated DDX1 constructs are shown and their respective solubility, stability and ATPase activity (measured in a coupled assay) is indicated by "+" or "-" (yes or no). The hydrodynamic radius and the molecular weight as determined via DLS are shown on the right.

construct	length	solubility	stability	ATPase activity	hydrodynamic radius and mol. weight (according to DLS)
DDX1 full-length	1-740	+	-	n.a.	n.a.
DDX1-728	1-728	+	+	+	3.96 nm/85.8 kDa (see section 3.3.1)
DDX1-694	1-694	+	+	+	4.32 nm/105.2 kDa
DDX1-674	1-674	+	+	+	4.48 nm/114.5 kDa
DDX1-655	1-655	+	+	-	4.14 nm/95.2 kDa
DDX1-648	1-648	+	+	-	3.93 nm/84.2 kDa
DDX1-610	1-610	+	+	-	3.95 nm/85.1 kDa

Only a small number of N-terminal truncated species were identified in the MALDI-MS PMF analysis of the DDX1 fragments. They mostly lacked the entire N-terminal region of the polypeptide chain up until the SPRY domain. N-terminal truncated species were considered to be non-functional, since they lack the catalytically important Q- and Walker A motifs.

All C-terminally truncated protein constructs that were soluble and stable, could be purified by an identical protocol and showed similar oligomerization properties (**Table 3.1**). However, the three most severely truncated constructs were found to be ATP hydrolysis deficient (only assessed qualitatively, see section 2.4.4). The longest, but still stable C-terminally truncated construct ended at residue 728, which is located at the C-terminus of the last predicted helix (**Figure 3.1.1**). Protein DDX1-728 that only lacks 12 residues when compared to full-length DDX1 could be purified to high homogeneity (**Figure 3.1.3**).

Results

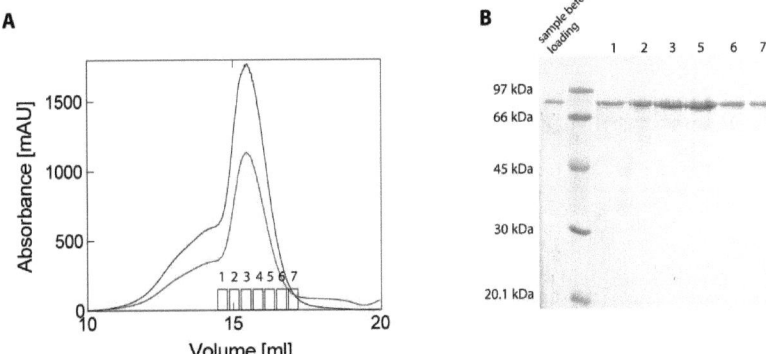

Figure 3.1.3 **Purification of construct DDX1-728 by size exclusion chromatography.** A construct of human DDX1 that was C-terminally truncated by only 12 amino acids could be expressed as a stable fragment in *E. coli* and purified to high homogeneity. **A**, Chromatogram from a gel-filtration/size-exclusion run on a Superose S6 column© (see material and methods) is shown. The absorbance at 280 nm is indicated by a blue line, absorbance at 260 nm with a red line, respectively. Fractions 1-7 were pooled and analysed are indicated below. **B**, 15 % (w/v) SDS-PAGE with the indicated elution fractions showed a single protein band after staining with Coomassie InstantBlue®.

The majority of recombinant DDX1-728 protein eluted from the Superose S6 column© at a volume, corresponding to monomeric species (**Figure 3.1.3**). Only a low amount of protein eluted prior to the main peak and these supposably aggregated fractions were discarded. Since DDX1-728 protein was used for all biochemical and biophysical studies on on human DDX1, "DDX1" always refers to construct DDX1-728 in the rest of the manuscript, unless explicitly stated.

3.1.3 Expression of the SPRY domain of DDX1

To identify stable domains within DDX1, limited proteolysis experiments were performed. The results of these experiments should guide the design of constructs that were used for crystallization experiments. Recombinant DDX1 protein (construct DDX-694) was digested with trypsin, chymotrypsin, thermolysin and proteinase K (see material and methods, section 2.3.8). Reactions were quenched at defined time points, reaction products were separated by 15 % (w/v) SDS-PAGE (**Figure 3.1.4 a** and **suppl. Figures 7.3.7** and **7.3.8**) and protein bands were excised from the gel and their identity was determined by MALDI-MS PMF.

Results

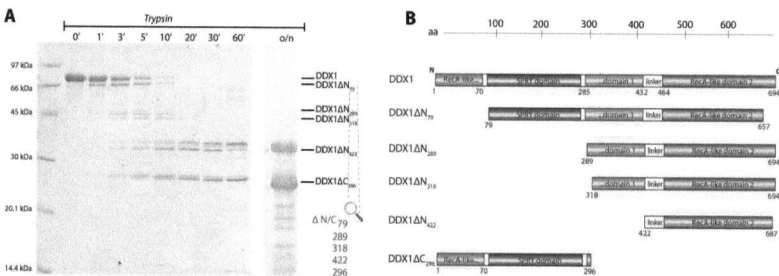

Figure 3.1.4 **Limited proteolysis of DDX1 to identify stable domains. A**, Construct DDX1-694 was digested with trypsin protease for the indicated time points (in min) or at 277 K over night (= o/n). Degradation bands were excised from the gel and identified via MALDI-MS PMF. Domain boundaries as derived from the detected peptides are indicated on the right side with ΔN and ΔC denoting N- and C-terminal truncations, respectively. Interestingly, most fragments were truncated directly at the boundaries of the SPRY domain. Fragment DDX1ΔN79 lacks all amino acids N-terminal to the SPRY domain. Fragments DDX1ΔN289 and DDX1ΔN318 lack the N-terminus and the SPRY domain. Fragment DDX1ΔN422 lacks RecA-like domain 1. Fragment DDX1ΔC296 lacks all amino acids downstream of the SPRY domain. Note that the protein construct used for the experiment, DDX1-694, is already C-terminally truncated. **B**, Domain boundaries as identified by MALDI-MS PMF are indicated on a representation of DDX1-694. Note that domain boundaries only correspond to the peptides that were detected and do not necessarily represent the actual protein fragments.

Interestingly, most proteolysis fragments did correspond to truncations directly at the boundaries of the SPRY domain (**Figure 3.1.4 b**). Fragment DDX1ΔN$_{79}$ was cleaved directly at the beginning of the SPRY domain, whereas fragments DDX1ΔN$_{289}$, DDX1ΔN$_{318}$ and DDX1ΔC$_{296}$ were cleaved at the end of the SPRY domain. These results showed that the SPRY domain per se is a stable fragment. Furthermore, when the products of a 24 h digest of DDX1-728 with trypsin were loaded on a Ni^{2+}-NTA affinity column, fragments corresponding to the SPRY domain were found in the flow-through, whereas RecA-like domains 1 and 2 were retained (**suppl. Figure 7.3.8**). This shows that the SPRY domain does neither interact with the Ni^{2+}-NTA matrix nor with the RecA-like domain fragments (that bind to the column even without a 6xHis-Tag), at least not under the given conditions.

Since the SPRY domain was proven to be highly protease resistant, constructs for its expression were designed. Domain boundaries were inferred from the MALDI-MS PMF results and refined based on sequence alignment of the entire sequence of human DDX1 with other helicases (also see **Figure 3.2.12**). Initially, the entire region that corresponds to the DDX1 specific insertion (amino acids 72 to 283) was cloned and used for expression (**suppl. Figure 7.1.3**). Secondary structure prediction by PSIPRED[177] for the SPRY domain suggested that N- and C-terminal residues do not adopt a defined structure. At closer inspection of the sequence alignment (**Figure 3.1.1**) these N- and C-terminal regions may constitute linker regions that connect the SPRY domain with the helicase core. This fits well with the previous results from limited proteolysis on DDX1-694 (**Figure 3.1.4**) that showed several cleavage sites in the putative linker regions, indicative of flexibility. Separate limited proteolysis experiments on the SPRY domain protein were performed (**suppl. Figure 7.1.4**) and based on the results, additional SPRY domain constructs with N- and C-terminal truncations were designed. All truncated SPRY domain constructs are summarized in **Table 2.1** (section 2.2.14, materials and methods). Constructs were tested for expression and protein solubility. Constructs that were stable and homogeneous in solution were used in crystallization screening (**see section 3.2.1**).

3.1.4 Expression of components of the HSPC117 complex and pull-down experiments with DDX1

CGI-99 and Fam98b

The HSPC117 tRNA ligation complex is a five component system[114, 118] (**see Introduction, Figure 1.10**). The two complex components CGI-99 and Fam98b[108] were cloned with an N-terminal 6xHis-tag and expressed in *E. coli* (**see material and methods, Section 2.3.5**). Cell lysates of both constructs were pooled together with DDX1-728 expressing *E. coli* lysate and purified via Ni^{2+}-NTA affinity chromatography.

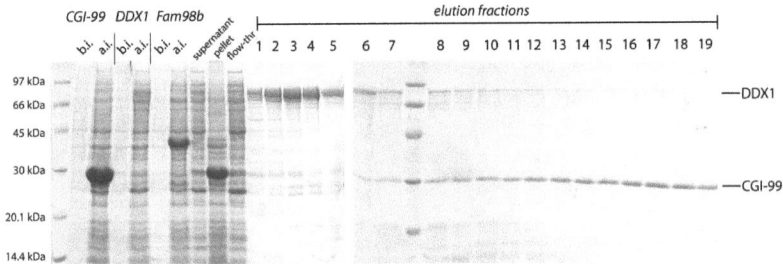

Figure 3.1.5 **HisTrap run with HPSPC117 complex components.** Proteins CGI-99, DDX1-728 and Fam98b were expressed in *E.coli* with an N-terminal 6xHis-tag. Bacteria cells before induction (b.i.) and after induction (a.i.) of the respective construct are shown. Bacteria lysates were pooled together and loaded on a HisTrap (Ni^{2+}-NTA affinity) column. Elution fractions of an Imidazole gradient are shown.

A significant overexpression of CGI-99 and Fam98b was observed on SDS-PAGE (**Figure 3.1.5**). However, only CGI-99 could be found in the eluate fractions. The majority of CGI-99 did elute later than DDX1, thus the proteins do most likely not form a stable complex under the purification conditions. CGI-99 could be further purified to high homogeneity using anion-exchange and gel-filtration chromatography. Upon concentrating above ~2 mg/ml (= 75 µM) it precipitated and thus, was unsuitable for crystallization experiments.

Results

ASW

The fourth HSPC117 complex component, the protein ASW[119] was cloned and expressed in E. coli with an N-terminal 6xHis-tag (see material and methods). The protein was found in the soluble fraction and could be purified to good homogeneity. Upon concentrating above ~5 mg/ml (= 195 µM), it did start to precipitate. In a gel-filtration chromatography run, most protein was found in the void-volume of the column, indicative of aggregation. To investigate the interaction of ASW with DDX1, the 6xHis-tag was cleaved off by thrombin digestion, undigested protein was removed via re-chromatography, and ASW protein was loaded on a His-trap column, onto which DDX1-728 was immobilized (**Figure 3.1.6**).

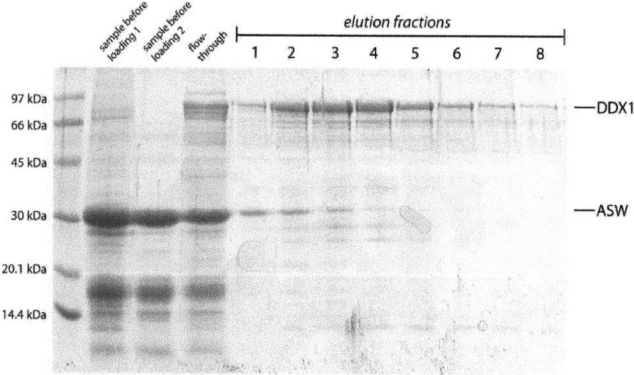

Figure 3.1.6 **HisTrap run with DDX1-728 and ASW.** A HisTrap (Ni^{2+}-NTA affinity) column was pre-loaded with DDX1-728. ASW protein (not containing a His-tag) was loaded on the column with immobilized DDX1 and proteins were eluted by an Imidazole gradient.

The majority of ASW protein was found in the flow-through. Only small amounts of ASW did co-elute with DDX1-728 in the first few fractions (**Figure 3.1.6**), which can probably be attributed to unspecific binding to the column matrix. Thus, under the conditions of the experiment, ASW and DDX1 do not form a stable complex.

3.1.5 Summary of the recombinant protein expression

The results of this section demonstrated that it is possible to recombinantly express human DDX1 in E. coli cells. However, determining the identity of protein fragments obtained from full-length DDX1 by MALDI-MS PMF revealed several C-terminally truncated species. Based on these observations, protein constructs with similar C-terminal truncations were designed. Dependent on the extent of the truncation, some protein constructs were ATPase deficient, but did otherwise not differ in their biophysical properties and could be purified by an identical protocol. Interestingly, despite the presence of harshly truncated proteins in the full-length DDX1 preparation, truncation by 12 residues at the C-terminus was sufficient to yield the stable protein construct DDX1-728.

Limited proteolysis experiments were performed to detect domain boundaries within DDX1 and the SPRY domain was identified as a protease resistant fragment. Since secondary structure prediction and the determined cleavage sites indicated unstructured termini, different truncated variants of this SPRY domain construct were designed.

The protein constructs, described in this section form the basis for experimental characterization of DDX1, described in paragraphs 3.2 and 3.3.

3.2 Structural studies

This chapter describes the structural studies on DDX1 that were performed using X-ray crystallography. Proteins of the SPRY domain of DDX1 were used in crystallization experiments, which resulted in crystals under a few conditions. Several rounds of optimization were required that finally led to the formation of well diffracting crystals. A number of different crystal forms were obtained when proteins of different SPRY domain constructs were screened. Proteins of a construct that was lacking the 6xHis-tag yielded crystals that diffracted up to 2 Å resolution. Diffraction data were phased by MR and a three dimensional structure of the SPRY domain was determined at near atomic resolution. In the structure the SPRY domain adopts the β-sandwich fold of a modular insertion domain, typical for SPRY domains. Most importantly, the structural details were used to investigate a putative function of the SPRY domain as a protein-protein or RNA-interaction platform. The potential residues on the protein surface, that could be involved in such a function were identified. Crystallization experiments were also performed using the entire DDX1 helicase, but the protein was recalcitrant to crystallize, even in the presence of ATP-analogs and RNA. Finally, based on the conserved helicase core structure and the SPRY domain structure, a homology model of DDX1 was constructed.

Structure of the SPRY domain

3.2.1 Screening for well-diffracting SPRY crystals

The SPRY domain is the exclusive structural feature that distinguishes DDX1 from the canonical helicase structure[92], which has been determined for other DEAD-box proteins[30, 34, 62]. To gain insight in this unique domain insertion, recombinant protein material of the separated domain was produced for structural studies. Several soluble SPRY protein constructs that differ in the length of the N- and C-termini were obtained (**Table 3.2**). Proteins were tested for their suitability for crystallization experiments by characterizing their thermal stability in CD-unfolding and sample homogeneity via DLS measurements (**Figure 3.2.1**).

Results

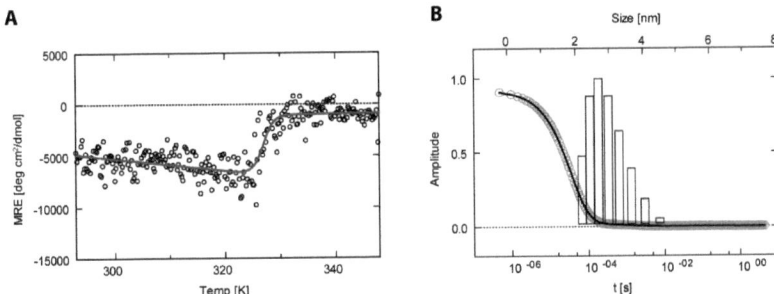

Figure 3.2.1 **Stability and homogeneity of SPRY crystallization construct.** The biophysical characterization of SPRY domain constructs is shown exemplarily for construct 'SPRY_72-283' that was used to solve the molecular structure. **A**, CD melting curve of SPRY_72-283 monitored at 222 nm and fitted to a two-state-unfolding process yields a T_m of 326.9 +/- 0.26 K. This melting temparature indicates a stable protein sample. CD values are in mean residue ellipticity = MRE. **B**, Data from DLS measurement with SPRY_72-283. The combined autocorrelation function (grey circles) is shown on the lower x-axis and a fit to this data is depicted as a black line. The upper x-axis shows the distribution of the hydrodynamic radii by relative mass (amplitude of each bar indicates the percentage of the total mass of the sample) as obtained from the fit. The fit did yield a peak in the mass-distribution of hydrodynamic radii at 2.89 nm (corresponding to a molecular weight of 40.8 kDa) and a peak width of 18.0 % RSD, which indicates a high degree of sample homogeneity.

These SPRY domain proteins were subjected to sparse-matrix screening experiments in 96-well sitting drop-format (JCSG core suite, Qiagen) and a few conditions were found in which growth of small protein crystals was observed. Larger crystals were obtained by fine screening with increasing protein- and reservoir volumes (in a 24-well format) and by streak-seeding into freshly prepared hanging-drop setups.

Table 3.2 **Different SPRY protein constructs used to determine the structure of the SPRY domain**
The solubility of the proteins and their tendency to form crystals is indicated by "+" or "-" (yes or no). For measured protein crystals, space group, unit cell parameters and lower diffraction limit are given. "n.a." means not applicable, "ΔTag" means the 6xHis-Tag was removed by thrombin digestion.

construct	solubility	crystals	space group	unit cell parameters	best-diffraction (in Å)
SPRY_72-283	+	+	$P2_12_12_1$	a = 43.73, b = 75.64, c = 122.73, α = β = γ = 90°	4.0
SPRY_72-283ΔTag	+	+	$P2_12_12_1$	a = 45.06, b = 76.14, c = 122.66, α = β = γ = 90°	2.0
SPRY_84-283	+	-	n.a.	n.a.	n.a.
SPRY_100-283	-	n.a.	n.a.	n.a.	n.a.
SPRY_72-261	+	-	n.a.	n.a.	n.a.
SPRY_84-261	+	+	P1	a = 35.47, b = 99.28, c = 136.36, α = 102.56°, β = 97.62° γ = 98.33°	2.7
SPRY_84-261 (SeMet)	+	+	$P2_1$	a = 37.14, b = 133.80, c = 35.96, α = γ = 90°, β = 118.26°	3.4
SPRY_100-261	-	n.a.	n.a.	n.a.	n.a.

Results

Largest crystals grew from construct SPRY_84-261 in 30 % (w/v) PEG 3000, 0.1 M CHES-NaOH pH 9.5 (**Figure 3.2.2 a**) and diffraction data were collected at the beamline X10SA (PXII) at the SLS in Villigen, Switzerland (see section 2.5.3). Crystals diffracted to 2.7 Å resolution and the symmetry belonged to the space-group $P1$. The Matthews coefficient[172] of 2.34 Å^3/Da suggested ten molecules per asymmetric unit. Due to the high number of expected molecules per asymmetric unit, it was impossible to obtain phases by molecular replacement. In order to get experimental phases, selenomethionine-labeled protein was expressed in *E. coli*. ESI-MS measurements confirmed that all seven methionines were substituted by selenomethionine (Se-Met). The Se-Met labeled SPRY_84-261 protein showed rapid formation of crystals under the same conditions as for the native SPRY_84-261 protein construct (**Figure 3.2.2 b**). Crystals diffracted to a resolution of 3.4 Å, however, the symmetry belonged to space-group $P2_1$. Due to low anomalous signal, phasing of diffraction data of the single anomalous dispersion experiment was not successful.

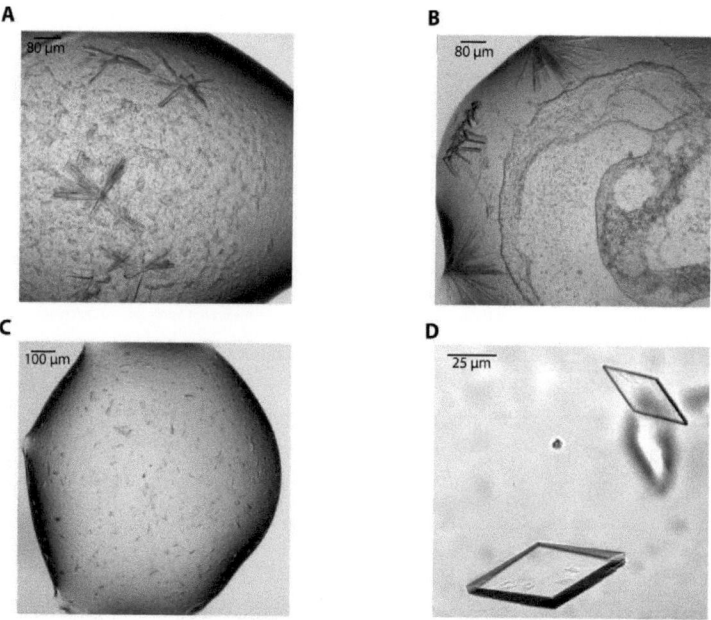

Figure 3.2.2 **Different crystal forms obtained with SPRY constructs.** Protein constructs of the SPRY domain crystallized in four different crystal forms, each form diffracting to a different resolution. **A**, crystals of construct SPRY_84-261 that grew in 30 % (w/v) PEG 3000, 0.1 M CHES pH 9.5. **B**, crystals of Se-Met version of SPRY_84-261 that grew in almost the same condition as the native protein, 30 % (w/v) PEG 3000, 0.1 M CHES pH 9.8. **C**, crystals of construct SPRY_72-283 that grew in 40 % (v/v) PEG 600, 0.1 M tri-Na-citrate pH 5.5. **D**, crystals of SPRY_72-283ΔTag (thrombin treated) that grew in 35 % (v/v) PEG 600, 0.1 M tri-Na-citrate pH 5.3.

Crystals that grew from the longest SPRY protein construct, SPRY_72-283 (**Figure 3.2.2 c**), were small and did diffract only to ~ 4.0 Å, but when the N-terminal 6xHis-Tag was removed (SPRY_72-283ΔTag), crystals of significantly improved diffraction quality were obtained (**Figure 3.2.2 d**). Most probably the unstructured and flexible 6xHis-Tag does not participate in crystal packing and its charge may hinder crystal growth.

3.2.2 Phasing and refinement of the SPRY structure

Crystals of construct SPRY_72-283ΔTag were used for structure determination of the SPRY domain. Spheroid crystals appeared after 3 days in 35 % (v/v) PEG 600, 0.1 M tri-Sodium-citrate pH 5.5 and grew as single crystals with typical dimensions of 140 x 90 x 40 µm^3 within five to six days at 297 K (**Figure 3.2.2 d**). For data collection, single crystals were harvested and soaked in reservoir solution for cryo-protection to prevent ice formation during flash-cooling in liquid nitrogen. Diffraction data were collected at the X10SA (PXII) beamline at the SLS in Villigen (Switzerland). Space group, unit cell parameters, and statistics of data collection are given in **Table 3.3**. Symmetry of the crystals of SPRY_72-283ΔTag belonged to the orthorhombic crystal system and the space group was $P2_12_12_1$. Diffraction data were processed with the XDS package[171]. Phases were obtained by molecular replacement (MR) with the software PHASER[170]. By using the BLAST service of the PDB website (rcsb.org) with the sequence of the DDX1-SPRY domain, the SPRY domain of human Ash2L[107] (PDB entry 3TOJ) with 23.8 % sequence identity was identified as the best search model. The automated molecular replacement mode of PHASER was used to search for two molecules within the asymmetric unit of the SPRY_72-283ΔTag crystals as indicated by a Matthews coefficient[172] of 2.01 Å^3/Da with a theoretical solvent content of 38.9 %. The search was carried out on data between 47 and 2.5 Å resolution. Based on the Z-score of the rotation- and translation function two solutions were found, but one of them was eliminated due to six clashes in the packing function. The final solution had a Z-score of 4.84 in the rotation function, a Z-score of 4.89 in the translation function and a log likelihood gain of 319 after final refinement. Phases were extended by a fourier-transform to the 2.0 Å resolution of the full dataset. Using the MR phases an electron density map was calculated and the model was built manually in Coot[46] with subsequent improvement in REFMAC[175] using restrained maximum likelihood refinement including TLS[176] refinement. Iterative cycles of refinement in REFMAC and model building in Coot led to a final model with an R-factor of 20.1 % and an R-free of 24.6 %. All refinement statistics are summarized in **Table 3.3**. Two molecules of SPRY (chains A and B) were found per asymmetric unit (**Figure 3.2.3**) with a solvent content of 34.5 %.

Table 3.3 **Statistics of data collection and refinement (molecular replacement).**
Values in parentheses are for the highest resolution shell.

PDB code	n.a.
X-ray source	beamline X10SA
Wavelength (Å)	1.06998
Space group	$P2_12_12_1$
Unit-cell parameters (Å,°)	a = 45.06, b = 76.14, c = 122.66,
	$\alpha = \beta = \gamma = 90°$
Observed reflections	367 329 (48 403)
Unique reflections	28 891 (3 815)
Resolution range (Å)	50-2.0 (2.1-2.0)
Redundancy	12.71 (12.69)
$<I/\sigma(I)>$	23.63 (6.52)
Completeness (%)	98.6 (97.6)
R_{meas}†(%)	7.9 (58.6)
Refinement statistics	
Resolution (Å)	47.78-2.0
No. of reflections (used in refinement)	27 446
No. of reflections (used for calculation of R_{free})	1 445
R_{work}/R_{free}‡ (%)	20.09/ 24.19
No. of non-H atoms	3154
Protein	3002
Water-molecules	152
Average B factors (Å^2)	31.05
Protein (chain A/B)	30.59/31.13
Water-molecules	34.70
R.m.s. deviations from ideal geometry	
Bond lengths (Å)	0.010
Bond angles (°)	1.300
Ramachandran plot	
Most favoured regions	90.6 %
Additional allowed regions	9.1 %
Generously allowed	0.3 %
Disallowed regions	0.0 %

† $R_{meas} = \Sigma_{hkl}\ [n/(n-1)]^{1/2}\ \Sigma_i\ |<I_{hkl}> - I_{hkl,i}|\ /\ \Sigma_{hkl}\Sigma_i\ I_{hkl,i}$, where $<I_{hkl}>$ is the mean intensity of symmetry-equivalent reflections and n is the redundancy.
‡ $R_{work} = \Sigma_{hkl}\ |F_{obs} - F_{calc}|\ /\ \Sigma_{hkl}\ F_{obs}$ (working set, no σ cut-off applied), R_{free} is the R value calculated for 5% of the data set not included in the refinement.

3.2.3 Overall structure of the SPRY domain

The structure of the complete SPRY domain of DDX1 was determined at a resolution of 2.0 Å (**Figure 3.2.3**). This thesis thereby presents the first structural detail of the human RNA helicase DDX1. The protein construct used in the crystallization experiment consisted of a polypeptide chain from residue 72 to 283 (212 residues), however, a clear electron density was only present for residues 86-275 of chain A and 86-279 of chain B. The first fourteen residues at the N-terminus (aa 72-85) and the last five residues at the C-terminus (aa 279-283) could not be modeled. They might be disorderd in the crystal or might have been cleaved off during the course of the crystallization experiment.

The two models of the polypeptide chain differ in four residues at the C-terminus that are clearly present in chain B (depicted in green in **Figure 3.2.3**), but could not unambiguously be modeled in chain A, due to poorly defined electron density. These last four residues at the C-terminus (aa 276-279) form an additional β-strand (β16) and are held in place in chain B by an intramolecular interaction with strand β1, which stabilizes their conformation (**Figure 3.2.3 b**). This may be an artifact of crystal packing, since from sequence alignment these residues are not expected to be associated with the SPRY domain and in chain A the loop leading over to β16 (which here of course is not present) points away from the SPRY core (and strand β1). Chains A and B were superposed using secondary structure matching (SSM) in $Coot^{[46]}$ and show a core RMSD of 0.246 Å. Since this result demonstrates that chain A and B are almost identical, but for chain B more residues could be modeled, the following figures and discussion will focus on this chain.

Results

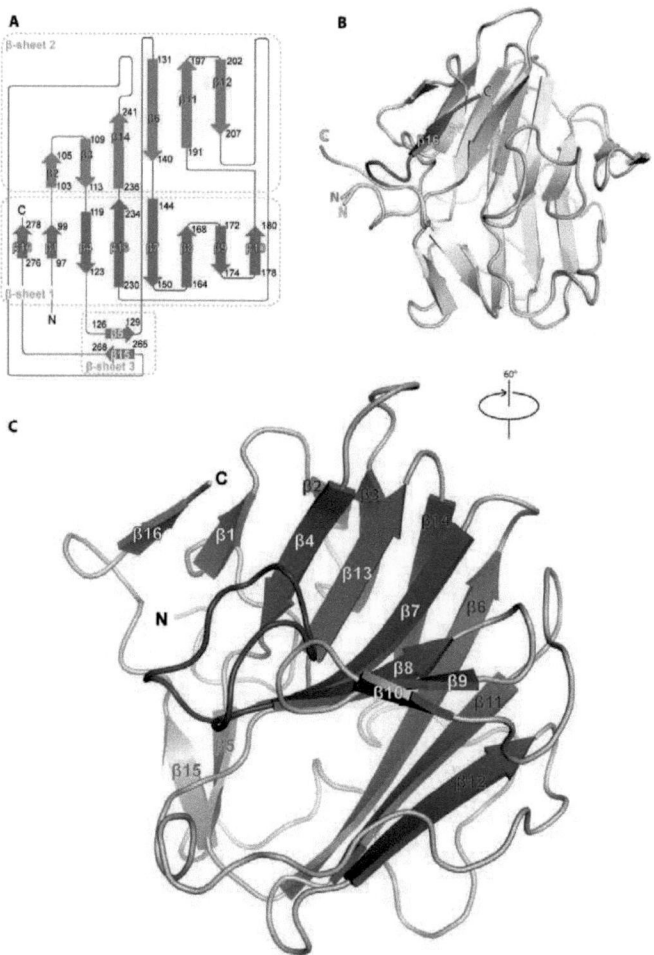

Figure 3.2.3 **Topology diagram and structure of the DDX1-SPRY domain**. **A**, a topology diagram of the SPRY domain is shown. The β-strands that form β-sheet 1 are depicted in blue (and labeled in white), the β-strands that form β-sheet 2 are depicted in red (and labeled in blue), the β-strands that form β-sheet 3 are depicted in green. **B**, the two copies of the SPRY domain in the asymmetric unit, chain A (depicted in yellow) and chain B (depicted in green) were superposed by secondary structure matching (SSM) in COOT and are shown in ribbon representation. They superpose with an RMSD of 0.246 Å and differ in four residues at the C-terminus that are present in chain B, but could not unambiguously be modeled in chain A. **C**, a ribbon representation of the β-sandwich fold of the SPRY domain is shown. β-sheet 1 is depicted in blue, β-sheet 2 in red and β-sheet 3 in green. Loop D is highlighted in purple. The different strands and N- and C-termini are labeled.

Results

The SPRY domain adopts a compact β-sandwich conformation. In contrast to other SPRY domains, that usually harbor α-helical regions at the N- and C-terminus[107], all secondary structure elements are β-strands. The overall β-sandwich fold is slightly distorted, forming a bowl-like platform. Two layers of concave shaped β-sheets stack together, in the following referred to β-sheet 1 and 2 (**Figure 3.2.3 c**). A third, much smaller β-sheet 3 forms a lid on the β-sandwich interspace. β-sheet 1 is composed of eight strands (β1, β4, β7, β8, β9, β10, β13 and β16; blue in Figure 3.2.2), β-sheet 2 is composed of six strands (β2, β3, β6, β11, β12 and β14; red in Figure 3.2.2) and β-sheet 3 only consists of two strands (β5 and β15; green in Figure 3.2.2). All β-strands that form the β-sandwich core fold are arranged in antiparallel configuration. Only strands β4/β5 and β15/β16 are oriented in a parallel direction (**Figure 3.2.3 c**). Short loops are connecting the β-strands on one edge of the β-sandwich, whereas on the other edge of the β-sandwich, where the β-sheet 3 lid is located, the loops are far more extended. Those extended loops include the long loop between β-strands β7 and β8, designated loop D (**purple in Figure 3.2.3 c**), which adopts a conserved conformation in other SPRY structures[107]. This loop lies in the bowl-like curvature of β-sheet 1. The conformation of loop D is locked by a salt bridge between Arg146 and Asp157 (2.7 Å) (**Figure 3.2.4 b**). Loop D does cover a hydrophobic patch on the concave side of the β-sandwich. An equivilant loop is found on the other, convex side of the β-sandwich (**suppl. Figure 7.1.5**). This loop connects β-strands β14 and β15 and also covers a hydrophobic patch on the β-sheet surface (but here on β-sheet 2). In the SPRY structure N- and C-termini are in spatial proximity, which is in line with a role as a modular insertion domain.

3.2.4 Interface between the two β-sheet layers

The SPRY domain adopts a compact β-sandwich fold. Salt bridges, hydrogen bonds and hydrophobic contacts contribute to the interaction at the interface of the two β-sheets and to conformational rigidity. Hydrogen bonds are formed between the oxygen atom of the side chain phenol of Tyr135 and the main chain amide of Gly148 (3.4 Å, **suppl. Figure 7.1.6 a**) as well as between side chain amide of Lys173 and the main chain carbonyl of Ala215 (2.6 Å, **suppl. Figure 7.1.6 b**), which stabilize the sandwich-fold. A salt bridge is formed between Glu184 and Lys207 (3.5 Å) and this ionic interaction locks the loop connecting β10 and β11 at the outer end of the domain, most distant from the termini (**Figure 3.2.4 c**).

Residues Cys139 and Cys145 can potentially form a disulfide-bond, which would bring β-strands β6 and β7 closer together, which are about 8.6 Å apart (C_α to C_α) in the structure (**Figure 3.2.4 d**). Formation of a disulfide-bond would also influence the conformation of the loop connecting both strands, which is one of the potential interaction loops (see below). The distance between the sulfur atoms of Cys139 and Cys145 is 4.3 Å (**Figure 3.2.4 d**). The protein was expressed under reducing conditions and thus, the potential disulfide-bond can not form.

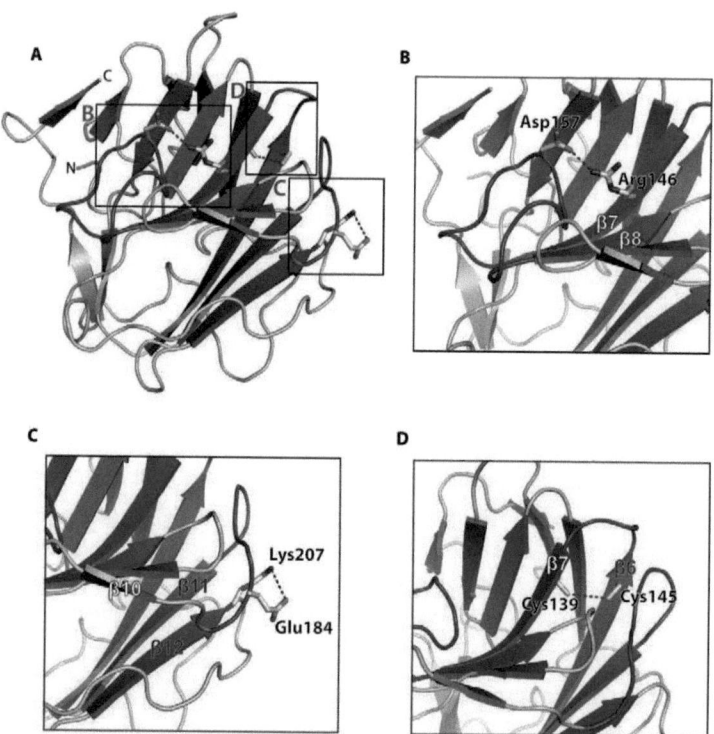

Figure 3.2.4 **Side chain interactions within the SPRY structure.** A, an overview of the SPRY domain structure is shown with windows that indicate the zoom area in the upcoming panels. B, close-up of loop D that lies in a bowl-like curvature and is stabilized by a salt bridge between Arg146 and Asp157. C, a salt bridge between Glu184 and Lys207 stabilizes the loop connecting β10 and β11. D, potential disulfide-bridge between Cys139 and Cys145. Residues are labeled in black and the loops that are stabilized in magenta. Color code for sheet 1, 2 and 3 and labeling of the β-strands are as in figure 3.2.3.

In addition to the ionic interactions, the β-sandwich fold is supported by an intramolecular hydrophobic core formed by the strands from β-sheet 1 with the opposite strands of β-sheet 2. Hydrophobic interactions involve residues from strands β2, β5, β6, β7, β8, β11, β12, β13 and β14 that stack together through Van der Waals contacts (**Figure 3.2.5**).

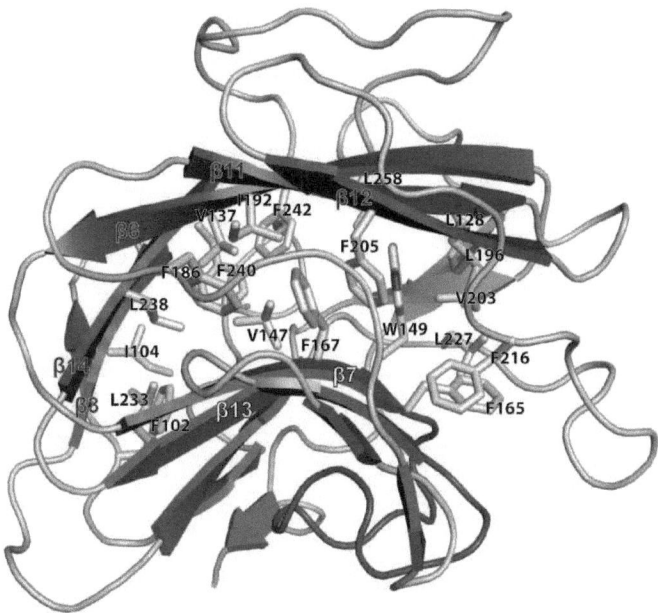

Figure 3.2.5 **Hydrophobic core in the SPRY domain.** The β-sandwich fold of the SPRY domain is stabilized by interaction of hydrophobic residues from the two β-sheets. Both sheets encompass an intramolecular hydrophobic core. Hydrophobic residues are depicted as stick models, labeled and colored in grey. β-sheet 1 is depicted in blue, β-sheet 2 in red and β-sheet 3 in green. The SPRY domain is rotated by 90° compared to figure 3.2.3 to better visualize the hydrophobic core.

3.2.5 Comparison of DDX1-SPRY to the structures of other SPRY domains

Since the first report of a NMR structure of the SPRY domain from murine SOCS box protein-2 in 2006[191], several crystal structures of mammalian SPRY domains have been determined. To detect and characterize particular features of the DDX1-SPRY domain, its structure was used for a structure-based sequence alignment with the other SPRY domain structures by the DALI web-server[192]. Several hits with high Z-score were obtained, showing that the same overall domain architecture is found in all SPRY structures. Amongst the best hits were the SPRY domains of human SPSB proteins 1, 2, 4[105], the SPSB orthologue protein GUSTAVUS from *D. melanogaster*[101] and human trithorax protein Ash2L[107]. With a Z-Score of 25.8 and an overall root mean squared deviation (rmsd) of 1.6 Å for the alignment of 170 residues, Ash2L-SPRY (PDB entry 3TOJ[107]) showed the highest structural similarity, despite bearing only 23.8 % sequence identity. The high structural similarity justifies the intial utilization of Ash2L-SPRY as a search model for MR in retrospect. It also proves that the BLAST routine of the PDB website is well suited to identify MR targets. The DALI superposition of DDX1-SPRY with the structure of the SPRY domain of Ash2L (**Figure 3.2.6 a**) was investigated in detail to characterize similarities and differences. Furthermore, the DALI superposition of DDX1-SPRY with the structure of the SPRY domain of GUSTAVUS in complex with 20-residue VASA peptide (PDB entry 2IHS[101], **Figure 3.2.6 b**) was investigated to compare the peptide interaction region. This peptide complexed structure of GUSTAVUS-SPRY showed similarity with a Z-score of 18.3 and 2.0 Å r.m.s. deviation for alignment of 152 residues (24 % sequence identity).

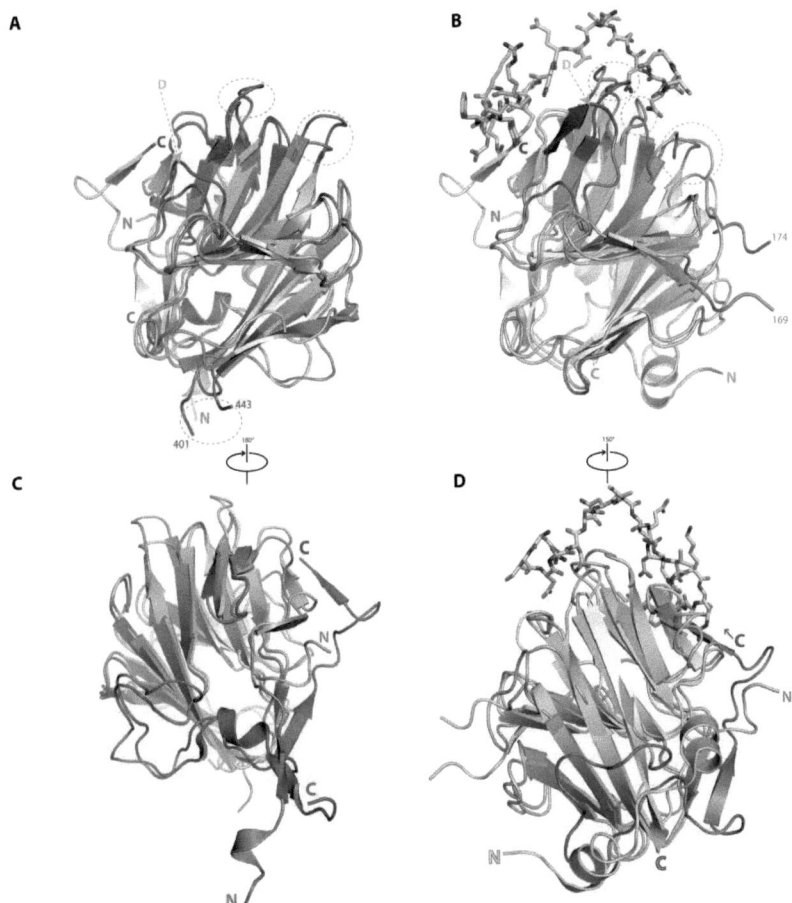

Figure 3.2.6 **Comparison of the DDX1-SPRY domain with other SPRY domains. A**, the SPRY domain of human DDX1 (depicted in green) was superposed with the SPRY domain of the protein Ash2L[107] (depicted in red, PDB entry 3TOJ). Regions that show most significant structural differences are indicated by more intense color shading. Interaction loops that differ in length and the long 44-residue loop, lacking in the Ash2L structure, are encircled. The loop D (as discussed in section 3.2.3) is marked. **B**, a superposition with the SPRY domain of the D. melanogaster protein GUSTAVUS[101] (depicted in yellow, PDB entry 2IHS) is shown. Color code and markings are as in A. The 20-residue VASA peptide is depicted in a stick representation (in blue). The region encompassing residues 169 to 174 is not resolved in the GUSTAVUS crystal structure. **C**, superposition as in A, but here the tadpole like extension of Ash2L-SPRY is highlighted that is not present in the DDX1-SPRY structure. C is rotated by 180° compared to A. **D**, superposition as in B, but here the N-terminus of GUSTAVUS-SPRY and the C-terminus of DDX1-SPRY are highlighted that both cover a hydrophobic patch on β-sheet 2. D is rotated by 150° compared to B.

SPRY domains have been reported to interact with other proteins via a common interaction surface A[100, 101, 105]. Interestingly, largest differences between DDX1-SPRY, Ash2L-SPRY and GUSTAVUS-SPRY are found in the loops that form surface A with r.m.s. deviations between 2.8 to 3.6 Å (**Figure 3.2.6 a** and **b**, grey circles). The loop connecting β3 and β4 is extended in DDX1-SPRY, when compared with Ash2L-SPRY and slightly tilted in comparison to GUSTAVUS-SPRY. The loop between β6 and β7 is shorter and retracted in contrast to both Ash2L-SPRY and GUSTAVUS-SPRY. Furthermore, the loop, connecting β13 and β14 is shorter in DDX1-SPRY in comparison to GUSTAVUS-SPRY. In GUSTAVUS-SPRY conserved loop D contributes to the interaction surface with the VASA peptide[101]. In DDX1-SPRY, however, loop D is shorter (by seven residues), far less extended and not part of surface A. Furthermore, a four residue β-strand (depicted in purple in **Figure 3.2.6 b**) seems to be an exclusive feature of loop D in GUSTAVUS-SPRY. In Ash2L-SPRY loop D is also shorter (when compared to GUSTAVUS-SPRY), but displays conformational differences in comparison to DDX1-SPRY.

In addition to the loops of surface A, further structural differences were detected when the DDX1 SPRY domain was compared to both other SPRY domain structures. Ash2L-SPRY contains a large 44-residue loop that connects β11 and β12, which was removed from the crystallization construct. The yeast Ash2L homolog Bre2 even contains two large loops, a 40 residue loop between β6 and β7 and a 120 residue loop between β11 and β12. In contrast to Ash2L and Bre2 human DDX1-SPRY lacks this extended loop. The corresponding loop between β11 and β12 is short and well resolved in the electron density (**Figure 3.2.6 a**). The largest loop in DDX1-SPRY is 23 residues between β14 and β15 at the C-terminus. In Ash2L-SPRY N- and C-terminus come together and form an additional β-sheet, which adds a tail to the compact SPRY domain core and generates an overall tadpole like structure[107] (**Figure 3.2.6 c**). In contrast, this convergence of N- and C-terminus is not observed in the DDX1-SPRY structure, where the C-terminus folds back on the SPRY core structure. Interestingly, GUSTAVUS-SPRY contains two helices at the N-terminus that show extensive hydrophobic interactions with β-sheet 2 and stabilize its curvature[100] (**Figure 3.2.6 d**). In contrast, in the SPRY domain of human DDX1 these interactions are taken over by the long loop in the C-terminal region that extends from β14 and crosses over strands β6, β11 and β12. Also in contrast to GUSTAVUS-SPRY, strand β4 is shorter and interrupted. After a short loop region, it continues into strand β5 that forms the separated β-sheet 3 (together with β15), which is not present in GUSTAVUS-SPRY.

To further investigate differences in peptide recognition motifs between the SPRY domain structures, which were identified in the DALI search, the sequences of the best hits were aligned and conservation values were mapped on the DDX1-SPRY surface (**Figure 3.2.7**).

Results

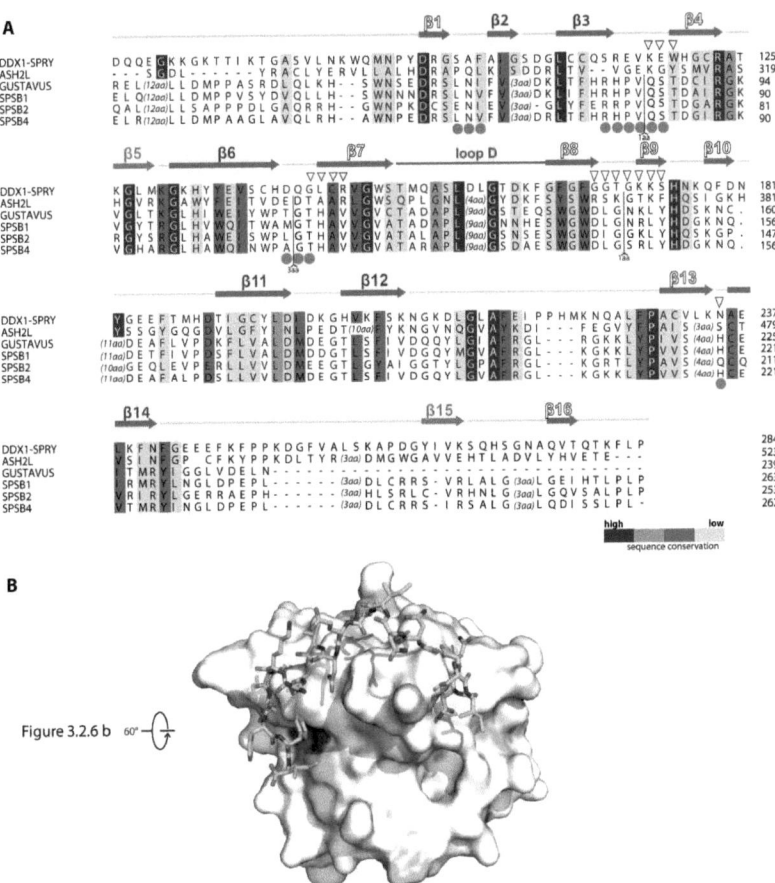

Figure 3.2.7 **Conservation of peptide interaction residues amongst SPRY domain structures.** The SPRY domain of human DDX1 is compared with other SPRY domains for which structures in complex with short peptides of the interaction partners are available[101, 105, 107]. **A**, the SPRY domain of human DDX1 is aligned with SPRY domains that show high structural similarity (as identified by a DALI[192] search). All proteins are of human origin, except for GUSTAVUS, which is a *D. melanogaster* protein. The degree of sequence conservation is indicated by color coding. Secondary structure is indicated above the sequence and β-strands are colored according to sheet 1 (=blue), sheet 2 (=red) and sheet 3 (=green). Loop D is indicated in purple. Residues of interaction loops that form surface A as identified in other SPRY structures[101] are indicated by grey circles. Residues that form a conserved positively charged surface patch in DDX1 homologs from different species (see below) are marked by triangles. **B**, the sequence conservation values are mapped on the surface of DDX1-SPRY. The short 20-residue VASA peptide from the structure of the GUSTAVUS SPRY domain[101] (PDB entry 2IHS) is superposed on the DDX1-SPRY surface. View is rotated by 60° compared to figure 3.2.6 b to better visualize the binding surface.

87

Noticeably, additionally to the conformational differences, least sequence conservation between Ash2L-SPRY and DDX1-SPRY is found in the loops connecting the strands β1 - β2, β3 - β4, β6 - β7 and β13 – β14, that form surface A (**Figure 3.2.7 a**). Investigation of the VASA peptide interaction in the complex structures revealed that only one of the residues (Val84 in GUSTAVUS/Val116 in DDX1) in the loops of surface A that mediate peptide-contact in GUSTAVUS-SPRY is conserved in DDX1-SPRY (**Figure 3.2.7 b**). Asn68, Arg81, Thr115 and Tyr133 (Ala101, Ser113, Gly143, Asp157 in DDX1) that form essential contacts to the SPRY interaction partner, not only in GUSTAVUS-SPRY[101], but also in the SPRY domains of human SOCS box proteins[105] are not found in DDX1-SPRY (**Figure 3.2.7 a**).

Overall the comparison of the DDX1-SPRY domain with the other SPRY domain structures, revealed profound conformational and compositional differences in the loop regions that mediate the contact with interaction partners in all other peptide complexed SPRY structures. Furthermore, the residues of the DDX1 specific extended surface A (see below) are not conserved amongst different mammalian SPRY domains (**Figure 3.2.7 a**).

3.2.6 The SPRY domain is a conserved interaction platform

To further characterize the role of the SPRY domain within DDX1, a sequence alignment of DDX1 orthologs of representative eukaryotic model organisms from primates to insects was used to map conservation values on the SPRY ribbon structure (**Figure 3.2.8**). Residues in the hydrophobic core that stabilize the β-sandwich fold were found to be either conserved or substituted with similar hydrophobic residues (grey diamonds in **Figure 3.2.8 a**). All β-strands are conserved, except for strand β15 of sheet 3 that shows a variation between hydrophobic and charged residues in different species. Residues of strand β5 that forms β-sheet 3 together with β15 are, however, conserved. Interestingly, strands β1, β4, β7, β8 and β9 of β-sheet 1 are virtually identical in all DDX1 orthologs, whereas the strands of β-sheet 2, mostly β2, β12 and β14 are less conserved (also see **suppl. Figure 7.1.7**). However, hydrophobic residues that contribute to conformational stability of the core are still preserved in less conserved strands. These residues include Ile104 of strand β2, Val203 and Phe205 of strand β12 and Leu238 of strand β14 (**Figure 3.2.8 a**). Least sequence conservation is found at the N- and C-termini of the SPRY domain. The termini correspond to the regions that connect the compact SPRY domain to the DDX1 helicase core, which again is well conserved across species (compare **Figure 3.1.1**).

Interestingly, the *D. melanogaster* and *C. elegans* DDX1 SPRY domains show differences in the polypeptide length and the loop connecting β5 to β6 is elongated by one residue in both species. The *C. elegans* protein displays additional length differences in the loop between β12 and β13 that is extended by two residues as well as in the loop between β14 and β15 that is extended by one residue. However, the loops that contain differences in the number of residues show low sequence conservation in the adjacent regions of the polypeptide chain and are part of the so called surface B (**see Figure 3.2.8 b**) for which no interactions have been described yet[100].

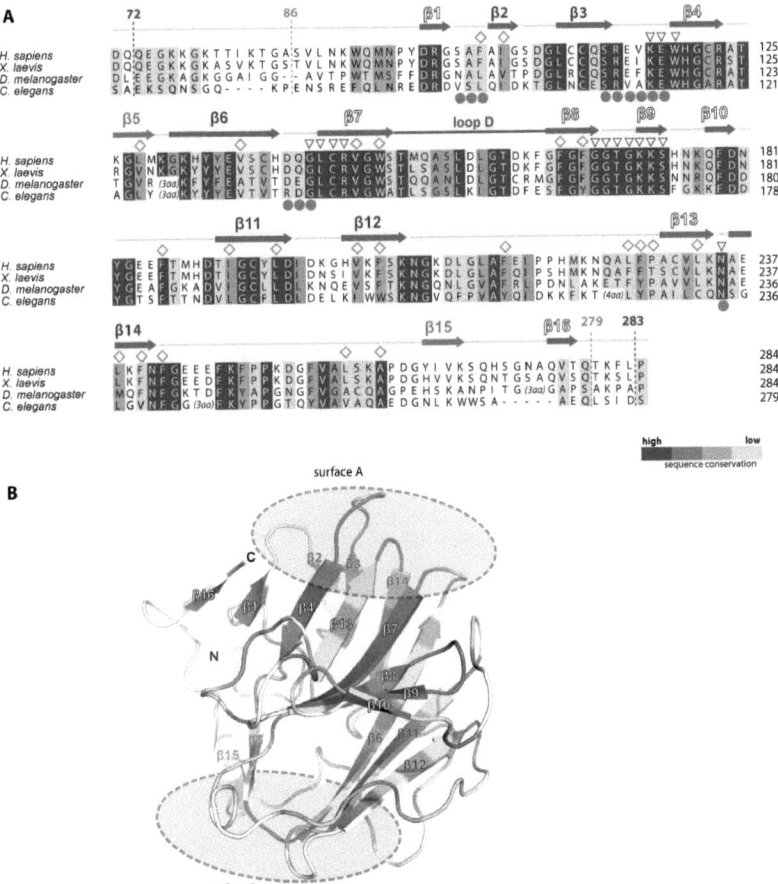

Figure 3.2.8 **Conservation of the SPRY domain amongst DDX1 from higher eukaryotes.. A**, The SPRY domain of human DDX1 is aligned with SPRY domains of DDX1 homologs from model organisms and the degree of sequence conservation is indicated by color coding. Secondary structure is indicated above the sequence and β-strands are colored according to β-sheet 1 (=blue), β-sheet 2 (=red) and β-sheet 3 (=green). Residues of the hydrophobic core that stabilize the domain fold are indicated by grey diamonds. Residues of interaction loops as identified in other SPRY structures[101] are indicated by grey circles. Residues that form a conserved positively charged surface patch are marked by black triangles (see below). Domain boundaries of the crystal structure (residues 86-279) are indicated in grey and domain boundaries of the crystallization construct are indicated in brown (residues 72-283). **B**, ribbon representation of the structure of the human DDX1-SPRY domain, colored according to sequence conservation. Surfaces A and B, formed by the loops that connect the β-strands are indicated by grey shading.

The residues that form the canonical interaction surface A in peptide complexed structures of other SPRY domains were found to be not conserved in the SPRY domain of DDX1 (see section 3.2.5 above). To find the potential location of a protein-protein interaction surface in DDX1-SPRY the sequence conservation on the protein surface was examined (**Figure 3.2.9**). A highly conserved patch could be identified that only partially overlaps with the location of surface A in other SPRY domains. In DDX1, this conserved patch is much more extended than in other SPRY domains and is mainly formed by the loop between β3 and β4, the upper parts of β13 and β7, the loop between β8 and β9 and the middle portion of β14 (**Figure 3.2.9 a**).

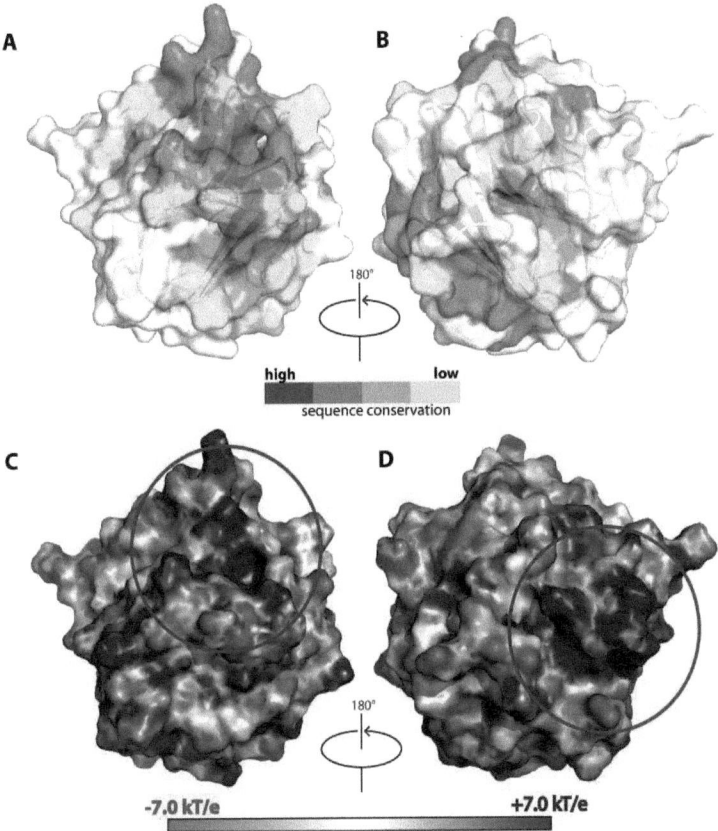

Figure 3.2.9 **Extended surface A, a conserved positively charged patch on the surface of the SPRY domain. A**, Sequence conservation (as obtained from sequence alignment -> see figure 3.2.8) is mapped on the surface of the DDX1-SPRY domain. The structure is oriented as in figure 3.2.6 and figure 3.2.8. The surface is partially transparent to show the peptide backbone. A large conserved patch is seen. This conserved surface is formed by the loop between β3 and β4, the upper parts of β13 and β7, the loop between β8 and β9 and the middle portion of β14. **B**, the same surface sequence conservation is shown, but the molecule is rotated by 180° around the y-axis compared to A. **C**, the electrostatic surface potential, calculated using APBS[183] is shown. The dielectric constant of the solvent was set to 80 kT/e. The contouring of the surface goes from +7 kT/e in blue to -7 kT/e in red. **D**, same surface potential as in C, but the molecule is rotated by 180° around the y-axis compared to C.

For mammalian SPRY domains it has been shown that mostly electrostatic interactions are established with protein partners bound to surface A[105]. The electrostatic surface potential of the DDX1-SPRY domain shows two elongated, positively charged patches that could serve as potential interaction regions. One of the positively charged surfaces is formed by residues of strands β7, β13 and the loop connecting β3 and β4 (**Figure 3.2.9 c**). Interestingly, this region is also conserved, suggesting that it may mediate the interaction with other protein factors or with RNA. To differentiate this conserved positively charged region from the canonical interaction surface A, it will be referred to as extended surface A.

The second positively charged surface patch is found at the N-terminus (**Figure 3.2.9 d**). It is most likely buried in the linker connecting the SPRY domain with the DDX1 helicase core and thus, probably not accessible for protein interactions. Furthermore, this charged region is not conserved (**Figure 3.2.9 b**).

Structure of DDX1

3.2.7 Crystallization trials with full-length protein

A number of structures of the canonical DEAD-box helicase core have been reported in recent years[31, 33, 36], making the SPRY domain the only novel structural feature of DDX1. Nevertheless, to obtain the complete picture and to determine the arrangement of the SPRY domain within the canonical core fold, crystallization experiments with full-length DDX1 were conducted. Crystallization trials using the apo-protein and co-crystallization with non-hydrolyzable ATP-analogs such as ATPγS and AppNHp were not successfull (**see appendix section 7.3.3**).
Most structures of DEAD-box proteins were determined in complex with RNA[37] (PDB entries 2DB3, 2HYI, 2J0S, 3I5X). In these structures RNA stabilizes the helicase core by binding across the two helicase domains, bridging them and bringing them close together[23, 24, 31, 35], ensuring a rigid scaffold required for crystallization. Co-crystallization experiments with different DDX1 constructs and different RNA oligonucleotides (as tested in EMSAs, see section 3.3.3) were performed (also with ternary complex of protein, ATPγS and RNA), but did not yield X-ray diffracting crystals (**see appendix section 7.3.4**).

3.2.8 Structural model of the entire DDX1 helicase including the SPRY domain

To get a picture of the entire DDX1 including all three domains, the SPRY domain and canonical RecA-like domains 1 and 2, a homology model of DDX1 was constructed. For this homology model only the basic helicase core of DDX1 without the C-terminal extension (residues 637-740) and without the SPRY domain and associated linkers (residues 70-284) was modeled. To find a suitable template for modelling, the PDB was searched for structures of DEAD-box proteins that harbor high sequence homology to DDX1. The highest BLAST[186] score of 443 (E-value 1.42352×10^{-43}) was obtained for the sequence of the apo structure of the *Methanococcus jannaschii* protein MjDEAD[33] (PDB entry 1HV8) with a sequence identity of 29.15 % to DDX1 (**Figure 3.2.10**). Using Swiss-MODEL[184], a structure of the DDX1 helicase core was assembled by sequentially mutating residues of MjDEAD to the corresponding residues in DDX1 and optimizing the geometry.

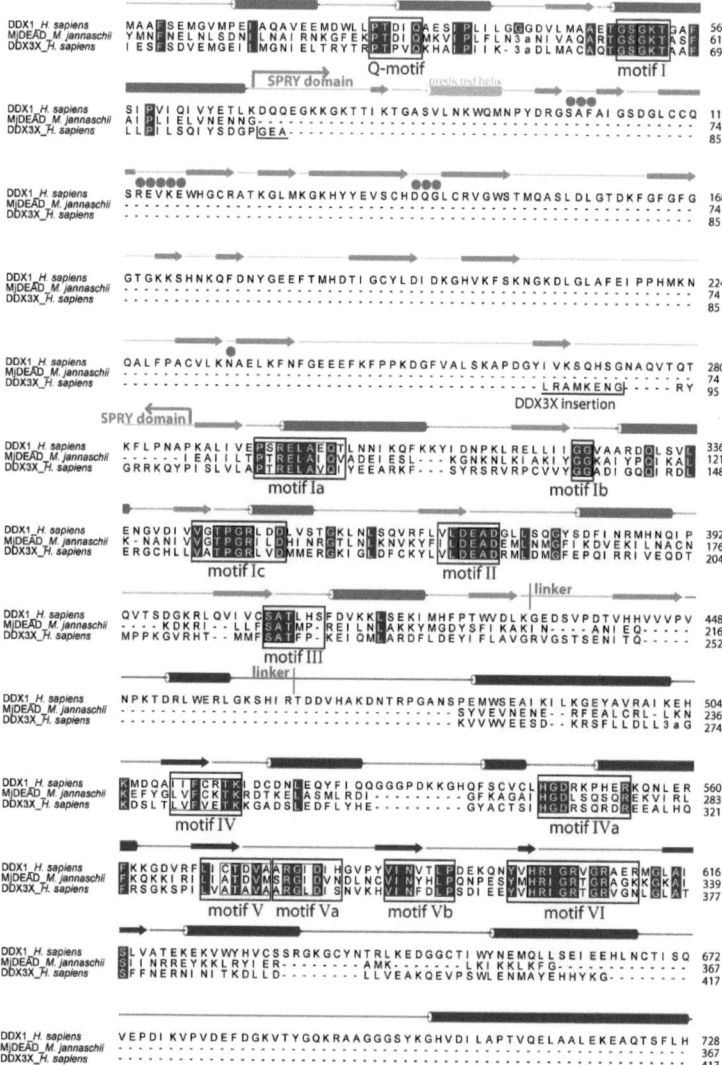

Figure 3.2.10 **Sequence alignment of DEAD-box proteins.** The DEAD-box proteins MjDEAD (from *M. jannaschii*) and DDX3X (from *H. sapiens*) were aligned with DDX1 (from *H. sapiens*). Residues that are identical are colored in green. Secondary structure as predicted for human DDX1 with PSIPRED[177] is indicated above the sequence and colored according to domain allocation. RecA-like domain 1 is colored in green, the SPRY domain is colored in blue and RecA-like domain 2 is colored in red. Conserved helicase sequence motifs (as explained in figure 1.3) are marked by blue bordering. The insertion in DDX3X that forms an elongated positively charged cavity is marked in brown.

Results

The SPRY domain was placed by an educated guess based on the structure of the human DEAD-box helicase DDX3X[185] (PDB entry 2I4I) (**Figure 3.2.10**). DDX3X only shows 21.04 % sequence identity and with a BLAST score of 257 (E-value 5.091x10^{-22}) only ranked 48 on the list of templates for homology modeling. However, similar to DDX1, DDX3X harbors an insertion domain, placed between conserved motifs I and Ia[185]. The insertion in DDX3X consists of ten residues that form an extended loop[185]. The DDX1-homology model was superposed with the structure of DDX3X using secondary structure matching (SSM)[46] and the SPRY domain was placed according to the position of the DDX3X-insertion (**Figure 3.2.11**). In this model the SPRY domain is oriented away from both the ATPase active site and RNA binding site on the surface between the two RecA-like helicase domains, suggesting that it neither interferes with ATP nor RNA binding (**Figure 3.2.11 b**). The DDX1 homology model and the structure of DDX3X represent helicases in the "open"-state. In general, upon binding of RNA substrate both RecA-like domains come together and form the "closed"-state as seen e.g. in the RNA-bound structures of yeast Mss116p[35], Dbp5[25] or Drosophila VASA[31].

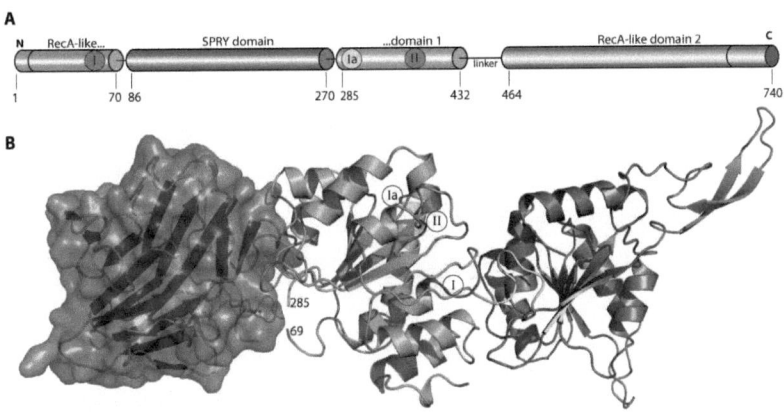

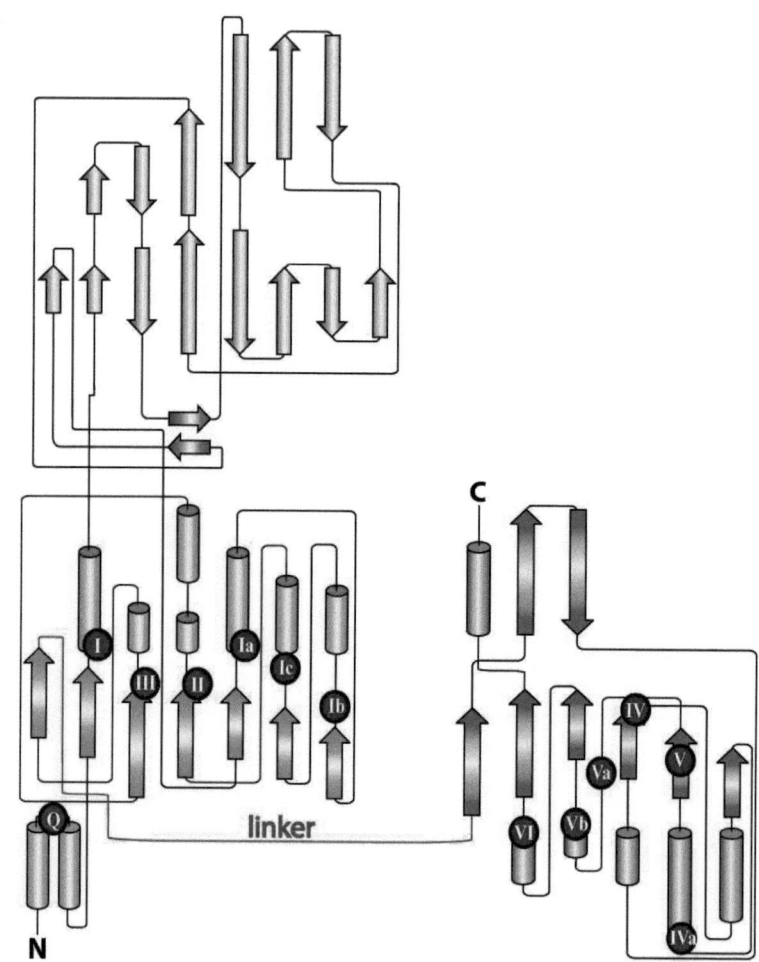

Figure 3.2.11 **Structural model of the entire DDX1 helicase.** A homology model of the helicase core of DDX1 was constructed based on the structure of MjDEAD (PDB entry 1HV8). The SPRY domain structure was placed in the model of the helicase core based on the position of the insertion in DDX3X (PDB entry 2I4I). **A**, Domain organization of DDX1 is shown schematically. The position of the conserved helicase motifs I (=Walker A), motif Ia and motif II (=Walker B) is shown by circles. **B**, homology model of DDX1 in the open conformation, based on the structure of MjDEAD, is shown. Color code is the same as in A. **C**, topology map of the DDX1 homology model, showing the position of the conserved helicase motifs.

3.2.9 Summary of the structural studies

The results of this section demonstrate how a divide and conquer approach succeeded in determining the structure of the distinctive feature of DDX1, the SPRY domain. The structure revealed three β-sheets, two of them stack together and form a β-sandwich and a third, small one forms a lid that closes over the β-sandwich core. A conserved surface patch was detected that is positively charged. This conserved patch, designated extended surface A, is not conserved in unrelated SPRY domains and does not match with the canonical interaction interface.

The structure of the SPRY domain was placed in a homology model of the DDX1 protein helicase core, based on the structure of the related human helicase DDX3X.

3.3 Functional characterization of DDX1

This chapter describes the functional characterization of DDX1 that was done using spectroscopic and biochemical methods. The functional results supplement the structural studies and helps to obtain a more complete picture of the unique DEAD-box protein DDX1. Initially, the recombinant protein-material was tested for its suitability for biophysical assays. Recombinant DDX1 was stable, monomeric and correctly folded. Affinities for nucleotides were determined and a profound preference for ADP over ATP was observed. Nucleotide affinity was modulated by RNA binding – cooperativity was most pronounced for ATP. Furthermore, RNA binding was characterized in gel-shift experiments and analysed in detail by a combined spectroscopic assay, which further supported cooperativity in RNA and ATP binding. DDX1 did display an intrinsic ATPase activity that can be stimulated by addition of RNA. Despite the ability to hydrolyze ATP, DDX1 was unable to unwind duplex RNA substrate in a helicase assay. Concluding from the affinity and ATPase measurements, RNA and ATP binding to DDX1 are tightly coupled.

Biophysical properties of recombinant DDX1

3.3.1 Folding and oligomerization of DDX1

Correct folding of recombinant DDX1 (**see section 3.1.2** on protein expression) was tested via CD spectroscopy (**Figure 3.3.1**). The spectrum of DDX1-728 showed the characteristic minima of a mixed α-helical/β-sheet protein, which is typical for the canonical RecA-like helicase fold[37]. However, the strong negative signal that is usually observed at shorter wavelengths (arising from α-helical secondary structure elements)[193] is attenuated by the SPRY domain insertion that consists solely of β-strands (**Figure 3.3.1 a**). Recorded thermal denaturation curves can be fitted to a two-state unfolding process yielding a melting temperature of T_m=326 K and an enthalpy at the transition midpoint of ΔH=646.6 kcal mol^{-1}, further supporting that the protein is folded correctly (**Figure 3.3.1 b**). Thermal denaturation of DDX1 was irreversible, as the protein was heavily aggregated in the cuvette after denaturation.

As a control in the biophysical measurements (**see section 3.3**) and to test the specificity of the ATP binding and the hydrolysis parameters, a mutant variant of DDX1-728, where the catalytically important Walker A lysine residue was mutated to alanine (K52A), was also studied. The K52A variant showed essentially the same behavior as WT protein during purification, gave the same CD-spectrum and the same melting point, indicative of correct folding (**Figure 3.3.1 a and b**).

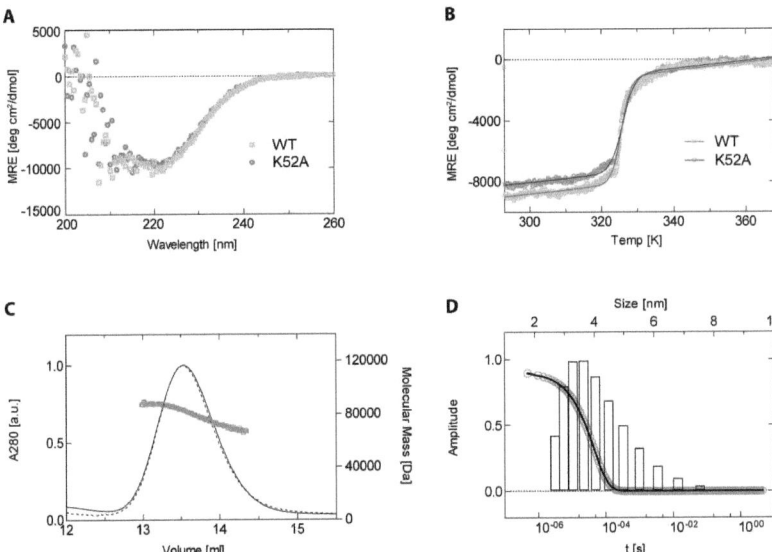

Figure 3.3.1 **Folding and Oligomerization state of DDX1. A**, CD spectra of DDX1-728 (=WT) and DDX1-728 Walker A K52A (=K52A) protein constructs, depicted in green and blue respectively are shown. CD signal measured with 5 µM protein was converted to mean-residue-ellipticity (MRE). **B**, temperature dependent unfolding curves as monitored by recording the CD signal at 222 nm. The data were fitted to a two-state-unfolding process. The fits yield a melting temperature T_m of 325.6 ± 0.04 K and enthalpy at the transition midpoint ΔH of 646.6 ± 16.9kcal mol^{-1} for DDX1-728, depicted in green and a T_m of 326 ± 0.08 K and an enthalpy ΔH of 520.5 ± 19.7kcal mol^{-1} for DDX-728 WalkerA K52A, depicted in blue. **C**, spectrum of a gel-filtration run with 50 µM DDX1-728 in 40 µl, coupled to a static-light-scattering detector. UV absorbance at 280 nm is shown on the left y-axis and a peak at an elution volume of 13.6 ml was observed, which corresponds to a molecular weight of ~ 80 kDa. On the right y-axis the dashed line represents the light scattering signal. The averaged molecular mass as determined by the MALS-data is plotted (in grey), which did correspond to 79.9+/-1.3 kDa. The **D**, data from dynamic light scattering measurements are shown. On the lower x-axis, the combined autocorrelation function from 20 x four second measurements is plotted in grey and a fit to the data in black. The upper x-axis shows the distribution of the hydrodynamic radii by relative mass (amplitude of each bar indicates the percentage of the total mass of the sample) as obtained from the fit. The peak at a hydrodynamic radius of 3.9 nm corresponds to a molecular weight of 85.8 kDa. Peak width is 24.4 % RSD, which corresponds to a homogeneous sample.

It has been reported that DDX1 forms dimers at high protein and low salt concentrations[135]. In fact recombinant DDX1 prepared during this thesis showed some aggregation (**see Figure 3.1.3**). Since formation of dimeric- or multimeric species complicates kinetic measurements, the formation of oligomers of DDX1 at the concentrations used in the biophysical assays was tested via static- and dynamic light-scattering (**see materials and methods section 2.4.3**). Even at concentrations of 50 µM, DDX1 did elute as a single peak from a size-exclusion column with a radius of gyration of 3.57 nm corresponding to a molecular weight of 79.9 +/- 1.3 kDa as determined from static light scattering data (**Figure 3.3.1 c**). DLS yielded a hydrodynamic radius of 3.96 nm and a corresponding molecular weight of 85.8 kDa (**Figure 3.3.1 d**). This shows that aggregated protein contaminations were efficiently separated from the main fraction by the gel-filtration chromatography during protein purification (**see Figure 3.1.3**), as theoretical mass of DDX1 is 82.4 kDa. Furthermore, both SLS and DLS measurements are in perfect agreement with DDX1 being present as a monomeric species at the concentrations used in the biophysical assays.

3.3.2 Characterization of conformational changes via limited proteolysis

Limited proteolysis has been successfully applied to map nucleotide binding dependent conformational changes of DEAD-box proteins[194]. To investigate the stabilization of individual domains of DDX1 in different nucleotide bound states, construct DDX1-728 was studied by limited proteolysis. The protein was digested with trypsin, chymotrypsin and thermolysin in the presence or in the absence of AppNHp (**Figure 3.3.2** and **suppl. Figure 7.3.7**). ADP was not included in the experiments, since it was not expected to induce any domain movement and did not show cooperative binding effects with RNA binding (**see section 3.3.6**).

Results

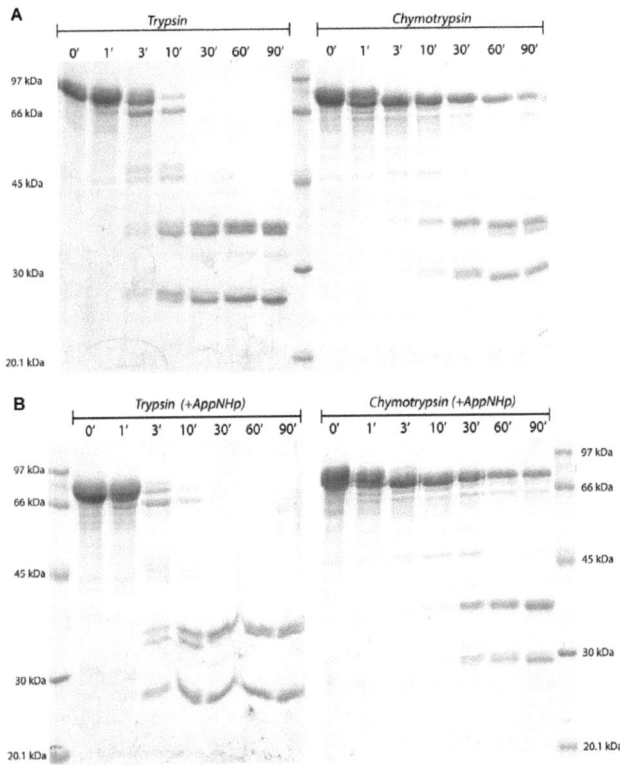

Figure 3.3.2 **Characterization of nucleotide dependent conformational changes by limited proteolysis.** Construct DDX1-728 was digested with trypsin and chymotrypsin for the indicated time points (in min) at 310 K. Degradation bands were excised from the gel and identified via MALDI-MS PMF. **A**, digest of a sample of apo DDX1-728 is shown. **B**, digest of a sample of DDX1-728 incubated with 10 mM AppNHp is shown.

Domain fragments generated by proteolysis were identical independent of the presence of AppNHp (**Figure 3.3.2**). Degradation bands were excised from the gel and further characterized by MALDI-MS PMF. They did correspond to the separated RecA-like domain 2 or to N- and C-terminally truncated DDX1 fragments. C-terminally truncated fragments were similar to constructs DDX1-655 and DDX1-694. N-terminally truncated fragments did lack all residues up to the SPRY domain.

Results

Nucleotide affinity of DDX1

3.3.3 RNA binding observed in gel-shift assays

DEAD-box helicases bind ssRNA through a set of conserved signature motifs on top of RecA-like domains 1 and 2[31]. An electrophoretic mobility shift assay (EMSA) was used to detect binding of short strands of RNA to DDX1. A 13mer RNA of random sequence, a 10mer polyU or 10mer polyA RNA were tested for binding (**see materials and methods, section 2.3.9**). The sequence of the 13mer RNA was randomized according to a protocol from Jankowsky and colleagues, that is classically used in studies of DEAD-box helicases[157]. To be able to visualize the oligonucleotides, FAM-labeled RNAs were used and illuminated on native gels by excitation of the FAM group at 302 nm. To optimize migration behavior of DDX1 (theoretical pI 6.9) in native gels, several buffer conditions and gel compositions had to be tested. Problems in the gel-migration of DDX1 have been reported before and in these experiments the complex of radioactively labeled RNA with DDX1 could not be visualized[135]. After optimization, it was possible to separate free- and protein-complexed RNA in 40 mM CHES (pH 10) buffered 5 % / 15 % (w/v) gradient gels at 277 K.

Results

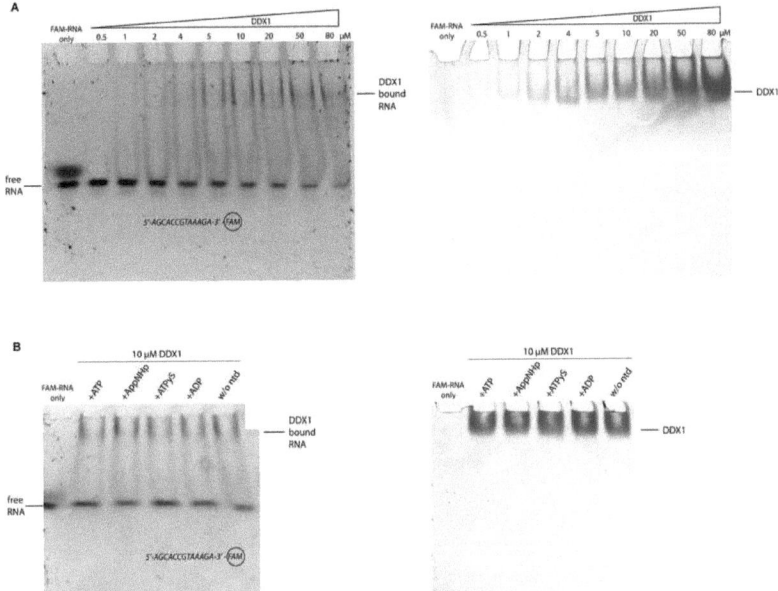

Figure 3.3.3 **RNA binding to DDX1 observed via gel-shift. A**, Binding of a 13-mer RNA strand, fluorescently labeled with a FAM-group at the 3'-end, to DDX1 was observed via a band-shift in a native gel. Increasing concentrations of DDX1-728 were incubated with 0.5 µM of labeled 13-mer RNA of indicated sequence for 20 min at 277 K. Reactions were separated on a 5 %/15 % (w/v) native acryl-amide-gel with CHES-NaOH pH 10 as running buffer. The FAM label was visualized by illumination at 302 nm. Protein was stained with Instant-blue to confirm co-migration with the RNA signal (see right side). **B**, same experiment as in A, but here 10 mM of different nucleotides (as indicated) were added to the binding reaction.

Binding of DDX1 to RNA was observed at relatively high protein concentrations (**Figure 3.3.3 a**). The position of the label on either the 3'- or 5'-end of the RNA substrate did not influence binding (**suppl. Figure 7.1.9**). All RNA species revealed a similar pattern of migration when incubated with DDX1, indicating that binding is sequence unspecific (**suppl. Figure 7.1.10**). Additional experiments were performed to investigate, whether the presence of nucleotides has an influence on DDX1 RNA binding or not. In the gel-based assay, no differences in the RNA binding affinity of DDX1 in the presence or absence of ADP, ATP or non-hydrolyzable ATP-analogs could be detected (**Figure 3.3.7 b**).

Of note, binding of DDX1 to RNA and DNA/RNA hybrid substrates was also observed in helicase assays (**see section 3.4**) in which a DDX1-substrate complex was shifted to the gel-pockets (**Figure 3.4.1**), whereas no binding and corresponding gel-shift could be observed for labeled, single-stranded DNA (**suppl. Figure 7.1.11**).

These gel-shift experiments only provide a qualititative description of the nucleotide affinities of DDX1. In these experiments, the binding equilibrium is disturbed and quantification of intensities is unreliable, therefore extraction of any binding parameters was not performed. To obtain a quantitive description of nucleotide binding to DDX1 and to extract correct values for the nucleotide affinities, fluorescence spectroscopy was used.

Cooperativity of RNA and ATP binding

3.3.4 Equilibrium titration of mant-nucleotides

The well-characterized fluorescent mant-labeled nucleotide analogs[195] were used to obtain a spectroscopically visible signal for nucleotide binding to DDX1. In spectroscopic experiments, fluorescence of the mant-group was excited directly by using UV-light at 356 nm and did increase upon binding to DDX1. This signal was used to characterize the affinity of DDX1 for ADP, ATP and RNA by a combination of different equilibrium titration experiments as described in the following.

Another way, which would allow to directly probe mant-nt binding is to use fluorescence resonance energy transfer between tryptophan residues of the protein and the nt mant-group, which has been described in the literature before[58, 196, 197]. DDX1 contains nine tryptophan residues in its amino acid sequence, one located close to the Q-motif in RecA-like domain 1, whereas the other tryptophan residues are located in the SPRY domain or in RecA-like domain 2. A strong tryptophan-fluorescence signal was obtained by excitation at 296 nm, but no energy-transfer to mant-nts could be observed. This means if DDX1 was titrated with mant-nts, it would be hard to differentiate between the increase in fluorescence caused by addition of the mant-nt and the one caused by mant-nt binding to the protein. Therefore, in all experiments in this thesis mant-nts were titrated with protein and not the other way round, which is not a problem due to the monomeric nature of DDX1.

For equilibrium titration experiments a constant concentration of mant-ADP was titrated with increasing amounts of DDX1 (**Figure 3.3.4 a**). The change in fluorescence due to binding can be fitted to the quadratic form of a single-site binding equation (Equation 8) resulting in a $K_{d,mantADP}$ = 0.12±0.02 µM for mant-ADP binding (**Table 3.4**). To test specificity of binding, the same titration was performed with the K52A mutant variant. For this variant full saturation of mant-ADP with protein and a plateau in the binding curve is not reached, even at high protein concentrations. Extrapolation of the binding curve to saturating conditions results in a $K_{d,mantADP}$ (K52A) = 31.4±4.9 µM, which represents more than 200 fold reduction in binding affinity (**Figure 3.3.4 b**).

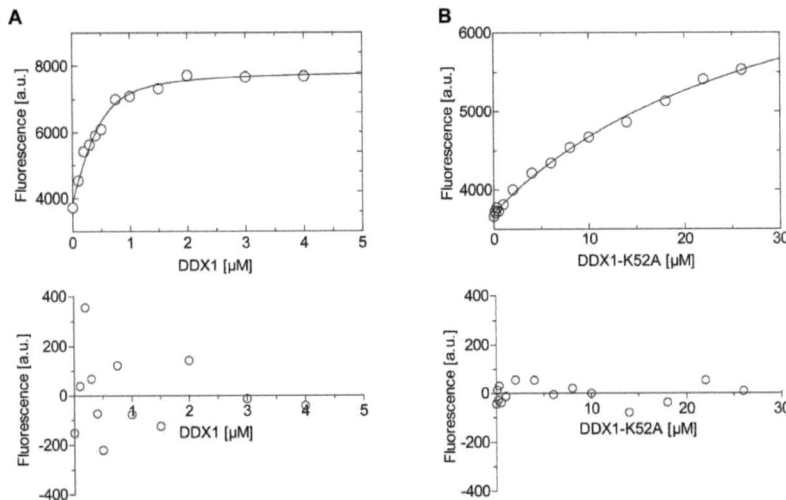

Figure 3.3.4 **DDX1 mant-ADP affinity assessed by equilibrium titrations.** **A**, in binding experiments 0.5 µM mant-ADP were titrated with DDX1 protein to obtain a $K_{d,mantADP}$. **B**, titration of 0.5 µM mant-ADP with DDX1 K52A variant is shown. To obtain affinity constants, binding data were fitted with the quadratic equation (Equation 8). Residuals from the fits (i.e. Δy between fit and actual data) are shown beneath the plots, where the y-axis represents 10 % of the binding amplitude. All parameters obtained from these experiments are shown in table 3.4.

To obtain binding affinities for unlabeled ADP, the mant-ADP DDX1 complex was pre-assembled and mant-ADP was displaced by titrating in excess ADP (**Figure 3.3.5 a**). Fluorescence data of the competition experiment can be fitted with the cubic equation[164] (Equation 9), resulting in a $K_{d,ADP}$ = 0.11±0.03 µM (**Table 3.4**). This value is close to the affinity for the mant-labeled ADP, indicating that the mant group has little influence on equilibrium binding, similar as reported for other systems[196]. Notably, when compared to the nucleoside-diphosphate affinities of other DEAD-box helicases[57, 58], K_d values for mant-ADP and ADP binding to DDX1 are smaller by at least two orders of magnitude.

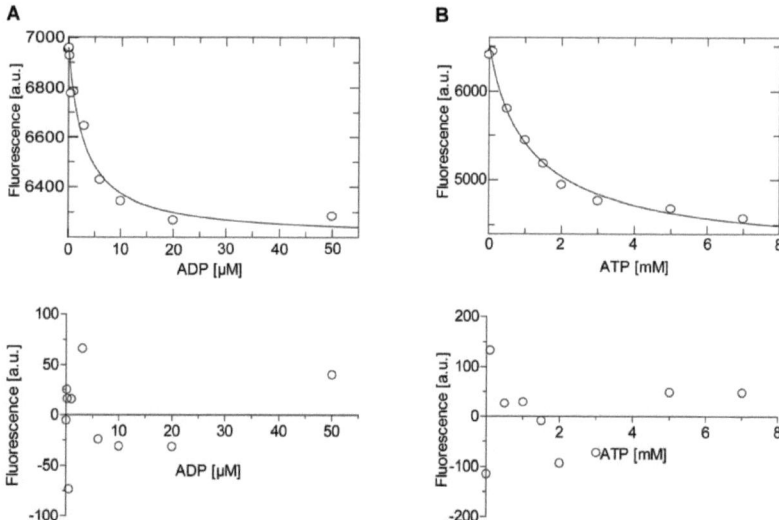

Figure 3.3.5 **DDX1 ADP and ATP affinities assessed by equilibrium titration in displacement experiments. A**, in displacement experiments a pre-equilibrated mixture of 3 μM mant-ADP and 0.5 μM DDX1 was titrated with unlabeled ADP to obtain an ADP affinity ($K_{d,ADP}$). **B**, a pre-equilibrated mixture of 0.5 μM mant-ADP and 2 μM DDX1 was titrated with ATP to obtain a $K_{d,ATP}$. Titration was performed in the presence of an ATP-regenerating system.
Data were fitted with a cubic binding equation (Equation 9). The mant-ADP affinity constant as determined beforehand was used as input parameter. Residuals from the fits (i.e. Δy between fit and actual data) are shown beneath the plots, where the y-axis represents 10 % of the binding amplitude. All kinetic parameters, obtained from these experiments are shown in table 3.4.

To probe binding of adenosine-triphosphates, the same experimental setup as before was used and mant-ATP was titrated with DDX1 protein. However, no significant increase of mant-fluorescence was observed in the titration experiment with 6 μM DDX1 as highest concentration (and 0.5 μM mant-ATP). Furthermore, hydrolysis of mant-ATP by DDX1 during the time-span of the titration experiment could potentially complicate measurements. Therefore, the experimental characterization was limited to the measurement of ATP affinity in competition with the mant-ADP DDX1 complex. To counteract ATP hydrolysis by DDX1, an ATP regeneration system, containing PK and PEP, was added to the reaction mixtures. PK has a very low turnover rate and affinity for mant-ADP (V_{max} of less than 2 % that of ADP), but rapidly converts ADP to ATP[195]. As expected from mant-ATP titrations, high concentrations of ATP were required to fully displace mant-ADP from DDX1 (**Figure 3.3.4 b**). A $K_{d,ATP}$ = 129±26 μM for ATP was obtained from a fit of the titration curves, which is within the range of what has been reported for other DEAD-box hélicases[57, 58, 75]. Though, this K_d value is higher by three orders of magnitude when compared to the affinity for ADP.

At closer inspection of the results of the characterization of ATP hydrolysis by DDX1 (**section 3.3.8**), this $K_{d,ATP}$ value has to be handled with care, since under the experimental conditions of the ATP equilibrium titrations, a fraction of enzyme may be in an ADP bound state. In this case, even in the presence of the ATP regeneration system, the titration with ATP would not measure a $K_{d,ATP}$, but a combination of ATP binding and the following steps (-> **see discussion 4.2.1**).

Since the fluorescent nucleotide mant-ADP exists in a mixture of the isomers 2'-mant-ADP and 3'-mant-ADP, titrations were repeated, using 2'-deoxy mant-ADP (mant-dADP) to exclude isomer specific effects. This variant of mant-ADP has the mant group attached to the 3' position of the ribose moiety and is unable to isomerize.
Whereas the apparent binding amplitudes (i.e. the signal) were different, all affinities were essentially the same (**suppl. Table 7.1** and **suppl. Figure 7.1.12**).

3.3.5 Transient kinetics of mant-nucleotide binding

The ADP affinity of DDX1, measured by equilibrium titrations, is unusually high, with a K_d smaller by at least two orders of magnitude when compared to values reported for other DEAD-box proteins[57, 58]. To validate this tight ADP binding, the affinity of DDX1 for Adenosine-diphosphates was also measured via transient kinetics. Pre-steady-state kinetic measurements were performed on a stopped-flow setup, to assess on- and off-rates for mant-nucleotide binding. Rapid mixing of excess mant-ADP with DDX1 leads to an increase in fluorescence. Fluorescence traces of experiments in which constant concentrations of DDX1 were mixed with different mant-ADP concentrations can be fitted to the sum of two exponential functions (**Figure 3.3.6**). The initial fast phase scales with mant-ADP concentration, whereas the second slow phase is constant throughout the experiments. Data analysis was limited to the initial fast phase and the second slow phase was excluded, because the amplitude is very small and it might reflect off-pathway isomerisation, small nucleotide contaminations in the protein preparations or mant-isoform specific interactions. The increase of fluorescence, described by the fast phase was interpreted as the mant-ADP binding process, because it shows a linear dependence on the mant-ADP concentration. The mant-ADP association rate constant was determined from the slope of the [mant-ADP] dependence of $k_{1,fast}$ giving an on-rate constant $k_{on,mantADP} = 0.95 \pm 0.12\ \mu M^{-1} s^{-1}$ for mant-ADP binding (**Table 3.4**). To obtain the off-rate constant, a chase experiment was performed, where DDX1 protein was pre-incubated with mant-ADP and subsequently mant-ADP was chased by mixing with excess amounts of unlabeled ADP. The time course of mant-ADP release can be described by a single exponential decay with a concentration-independent rate constant $k_{off,mantADP} = 0.15 \pm 0.0004\ s^{-1}$ (**Figure 3.3.6, Table 3.4**).

Thus, a resulting mant-ADP affinity of $K_{d,mantADP}$ = 0.160±0.021 µM was calculated from the transient kinetics measurements, which corresponds well to the affinity of $K_{d,mantADP}$ = 0.12 µM determined by equilibrium titrations (**see above section 3.3.3**).

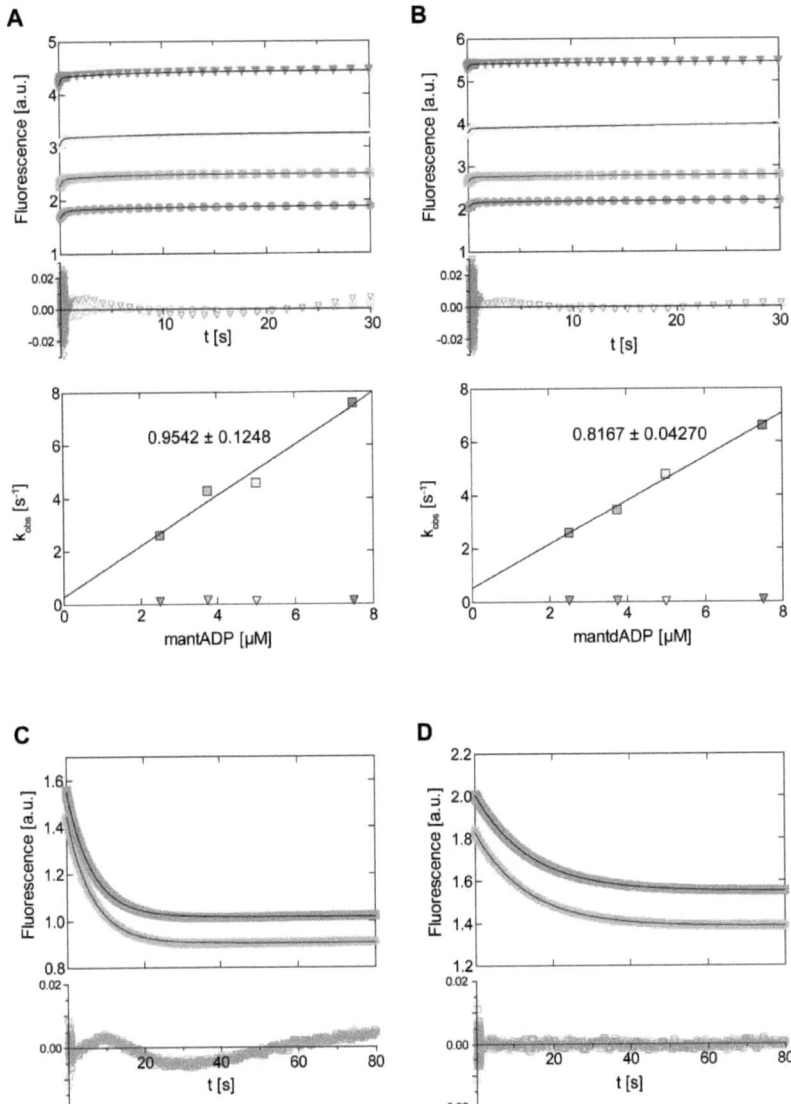

Figure 3.3.6 **DDX1 binding to mant-ADP and mant-dADP assessed by transient kinetics.** A, fluorescence traces obtained by rapid mixing of 2.5 (blue), 3.75 (green), 5 (yellow) and 7.5 µM (red) mant-ADP with 0.5 µM DDX1. Traces can be described by double exponential fits (black lines) and the corresponding rate constants $k_{1,fast}$ (squares) and k_2 (triangles) are plotted below. The fast rate constant $k_{1,fast}$ (squares) scales linearly with mant-ADP concentration and the slope (0.954 as indicated) was used as on-

rate for binding. **B**, same experimental setup as before, but here different concentrations of 2'-deoxy variant of mant-ADP (=mant-dADP) were rapidly mixed with DDX1. **C**, fluorescence traces of chase experiments, where an equilibrated solution of 0.5 µM mant-ADP and 0.5 µM DDX1 was rapidly mixed with either 250 (green) or 500 µM (blue) unlabeled ADP. Single exponential fits (black lines) gave the kinetic off-rates for mant-ADP dissociation. **D**, same experimental setup as in C, but here the complex of DDX1 with mant-dADP was chased by excess ADP.
Residuals from the fits (i.e. Δy between fit and actual data) are shown beneath the plots, where the y-axis represents 10 % of the binding amplitude. In the residual plots data points are depicted as hollow symbols.

As before, to exclude isomer specific effects and to test whether the second phase in the time course of mant-ADP binding is due to different affinities of DDX1 for 2'- or 3'-mant-ADP, the experiments were repeated using 2'-deoxy mant-ADP. Binding of mant-dADP is identical to mant-ADP, but the second slow phase is less pronounced (**Figure 3.3.6 b**). The mant-dADP concentration-dependent first fast phase gives a kinetic on-rate constant $k_{on,mantdADP} = 0.817 \pm 0.043$ µM^{-1}s^{-1}. Chase experiments with excess ADP give an off-rate $k_{off,mantdADP} = 0.074 \pm 0.0001$ s^{-1} (**Figure 3.3.6 d**). The corresponding affinity $K_{d,mantdADP} = 0.091 \pm 0.005$ µM corresponds well to the results from equilibrium titration measurements with mant-dADP ($K_{d,mantdADP} = 0.124$, compare **suppl. Table 7.1** and **suppl. Figure 7.1.12**).

3.3.6 RNA modulates the nucleotide affinity of DDX1

The established experimental setup of exploring DDX1 binding by an increase in mant-nucleotide fluorescence was used to determine affinities for ADP and ATP in the presence of RNA. To work under conditions of saturated RNA binding, 28 µM RNA was used in the titrations. Later experiments confirmed that under these concentrations DDX1 is saturated with RNA (**see section 3.3.7**). Three different RNAs were tested for their influence on DDX1 nucleotide binding (**Figure 3.3.7**). The 10mer polyA RNA and the 13mer RNA of random sequence[157] as mentioned before (**see section 3.3.3**) and a 20mer RNA of random sequence (**see methods for sequences, section 2.3.9**).

In the following text, to account for the ternary system of ATP and RNA binding to DDX1, the respective molecule that is present at saturating concentrations will be indicated in parentheses and subscript next to the given affinity constant.

Results

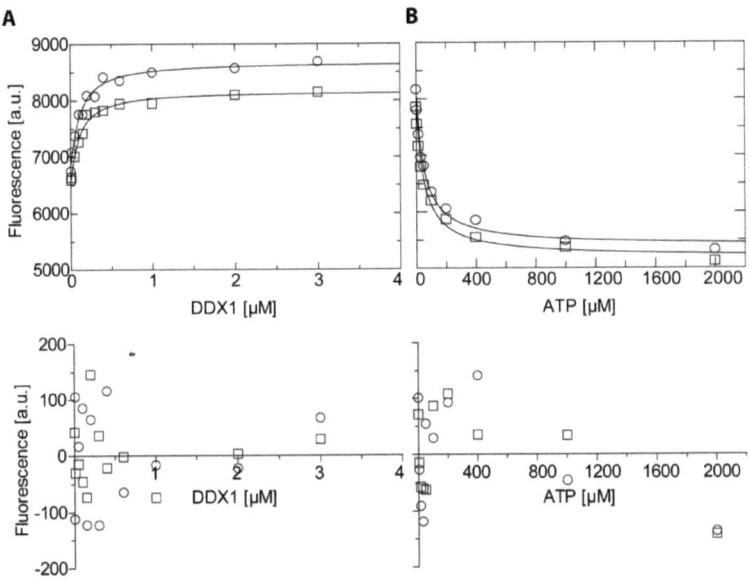

Figure 3.3.7 **RNA dependent nucleotide binding. A,** equilibrium titration experiments where 0.05 µM mant-dADP were titrated with DDX1 protein in the presence of saturating amounts (28 µM) of 10mer polyA RNA (circles) or 13mer random sequence RNA (squares) are shown. **B,** competition experiments in the presence of saturating amounts of the same RNAs are shown. For the experiments a complex of 0.2 µM mant-dADP and 1 µM DDX1 was titrated with ATP in the presence of an ATP-regeneration system. Titration data were fitted as described for figure 3.3.4 and corresponding rate constants are shown in table 3.4. Residuals from the fits (i.e. Δy between fit and actual data) are shown beneath the plots, where the y-axis represents 10 % of the binding amplitude.

The affinity for mant-dADP remained unchanged by any of the three tested RNAs (**Table 3.4**). In contrast, the affinity for ATP was increased ~ 20 fold from an RNA-free $K_{d,ATP}$ = 129±26 µM to a $K_{d(RNA),ATP}$ (RNA-saturated) = 5.0±0.8 µM (**Table 3.4**). The propensity to stimulate ATP binding was identical between different RNAs tested (**Figure 3.3.7 b**). Notably, even in the presence of RNA with increased ATP affinity, the apparent ATP affinity is still 50 times lower when compared to that for ADP. Again in consideration of the results from ATP hydrolysis, this $K_{d(RNA),ATP}$ value has to be interpreted with care, since it could be influenced by an enzyme that is mostly ADP bound under steady-state conditions (-> **see discussion 4.2.1**).

3.3.7 Quantification of RNA affinity by spectroscopic methods

A minimal binding scheme that describes the transitions between the different nucleotide bound states of DDX1 was defined (**Figure 3.3.8**). This binding scheme helps to interpret the relationship between the affinity constants, obtained in the spectroscopic measurements.

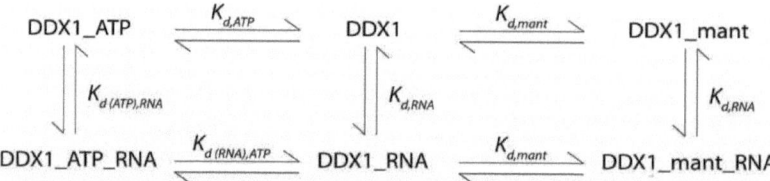

Figure 3.3.8 **Scheme for nucleotide binding to DDX1.** A minimal binding scheme for binding of ATP or mant-dADP (=mant) to the DDX1 helicase. This scheme was used to establish the relationships between the affinity constants and later on for global fitting of all titration data. The global fit according to this scheme allowed to obtain the affinities for RNA, $K_{d,RNA}$ and $K_{d\,(ATP),RNA}$.(see table 3.4 and figure 3.3.9).

So far, data for characterization of the dissociation constants of RNA in the minimal binding scheme are yet lacking. Initially, binding of RNA to DDX1 was observed in the EMSAs (see section 3.3.3) and different spectroscopic assays were used in attempts to quantify the binding affinity. However, all conventional approaches such as steady-state titration experiments with labeled RNA or labeled protein as well as FRET experiments with labeled Cys residues and via intrinsic Trp fluorescence or fluorescence anisotropy experiments failed to provide a signal for oligonucleotide binding to DDX1. Thus, the fluorescent mant-nucleotides were used in a more elaborate experimental setup. The complex of DDX1 with mant-dADP was pre-assembled and ATP was added at low concentrations, where it partially displaces the complex. In a next step, the experimental acquisition was started by titrating RNA that binds to DDX1, promotes further ATP binding (probably by inducing the transition to the "closed"-state of DDX1, see section 4.2.1) and thereby leads to further mant-dADP displacement. Following this experimental approach RNA binding could be monitored by a decrease in mant-dADP fluorescence. RNA titrations were performed at different ATP concentrations (**Figure 3.3.9**). Although these data do not instantly provide affinity values for RNA, together with the data from previous titrations they were included in a global fit analysis based on the minimal binding scheme (**Figure 3.3.8**). Global fitting was done by numeric iteration using the software Dynafit[198, 199] (**Figure 3.3.9** and **suppl. Figure 7.1.13**, see appendix **Section 7.8** for the script file).

Results

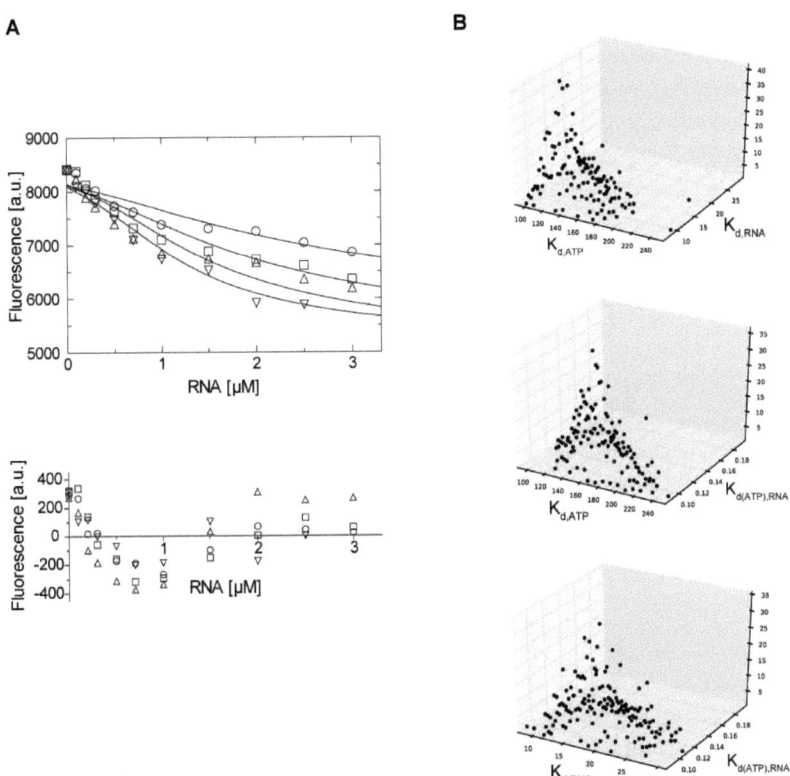

Figure 3.3.9 **Global fit of RNA titrations to obtain the RNA affinity of DDX1**. **A**, titration experiments to access RNA affinity are shown. Complexes of 1 µM DDX1, 0.2 µM mant-dADP and either 50 (circles), 100 (squares), 200 (triangles) or 400 µM (inverted triangles) ATP were titrated with 10mer-polyA-RNA. Titrations were conducted in the presence of an ATP-regeneration system. Fits that are shown were obtained by global numeric iteration using all previously measured titration data in the program Dynafit (BioKin Ltd., Watertown, MA). Residuals from the fits are shown beneath the plots. The affinities for RNA, $K_{d,RNA}$ and $K_{d\,(ATP),RNA}$ were obtained in the global fit and are shown in table 3.4. **B**, correlation between globally fitted values of selected affinity constants is shown. Affinity constants that are correlated are plotted on the x- and y-axis of each graph. The z-axis gives the number of counts for each individual correlation. This number of counts describes the convergence in the fit of the two parameters that were correlated. For the confidence intervals of individual kinetic parameters see suppl. figure 7.1.13.

The optimized kinetic paramters, obtained from the global fit analysis, finally allowed to access the dissociation constants for RNA. DDX1 showed an RNA affinity with a $K_{d,RNA}$ = 12.41±2.75 µM for RNA binding to nucleotide free- or mant-dADP bound DDX1. RNA binding to the ATP-bound form of DDX1 was determined to be tighter with a $K_{d(ATP),RNA}$ = 0.136±0.020 µM (**Figure 3.3.9**, Table 3.4). These K_d values for RNA binding contain important information on the difference between the RNA affinities of the ATP-bound form of DDX1 versus the nucleotide free form. The ratio between the two RNA affinities gives a ~90 times tighter binding of RNA to the DDX1-ATP-form.

Interestingly the K_d values for ATP binding to DDX1, obtained in the global fit, were slightly different to the values determined in the equilibrium titrations ($K_{d,ATP}$ = 129 µM, $K_{d,(RNA),ATP}$ = 5 µM, see **section 3.3.6**). Whereas ATP binding to the apo form of DDX1 was similar as before ($K_{d,ATP}$ = 124 µM), ATP binding to the RNA bound form of DDX1 was determined to be tighter ($K_{d,(RNA),ATP}$ = 1.4 µM, also see **suppl. table 7.2**). According to the minimal binding scheme (**Figure 3.3.8**), it follows that

$$K_{d,ATP} \cdot K_{d(ATP),RNA} = K_{d,RNA} \cdot K_{d(RNA),ATP} \quad i.e.$$

$$K_{d,RNA} = \frac{K_{d,ATP}}{K_{d(RNA),ATP}} \cdot K_{d(ATP),RNA} \qquad \text{[Equation 15]}$$

so the ratio of the ATP-affinities would also predict a ~ 90 times tighter binding of RNA to the DDX1-ATP-form, when using the K_d values from the global fit. When the ATP affinities, obtained from the equilibrium titrations are used, only a ~ 24 times tighter binding of RNA to the DDX1-ATP-form is calculated. The difference in the K_d values could be caused by small variations in the actual experimental concentrations that, however, have a large influence on the global fit. They could also result from a slightly incorrect determination of K_d for ATP by the equilibrium titration experiments as mentioned before.

Notably, when the same RNA titration experiments were performed in the presence of ADP instead of ATP (at ADP concentrations that lead to partial displacement of the mant-dADP – protein complex) the mant-fluorescence spectrum was unchanged (**suppl Figure 7.1.14**). This confirms the previous results from the equilibrium titrations in which RNA did not influence ADP binding (section 3.3.6) and further supports the assumption that ADP and RNA binding to DDX1 are not coupled.

Results

Table 3.4. nucleotide binding affinities of DDX1
Molecules in subscript and parentheses next to an affinity constant indicate the respective molecule that is present at saturating concentrations. The errors represent the standard error from the fits."+" means addition of a molecule, "-" means the experiment was performed without the molecule.
* please note that the K_d values for ATP measured via equilibrium titration have to be carefully interpreted, since - in conclusion from the ATPase data - they might be influenced by a fraction of DDX1 that is always ADP bound (see discussion section 4.2.1)

mant-ADP binding			
$K_{d,mantADP}$	0.124	± 0.017	µM
$K_{d,mantADP}$ (K52A)	31.4	± 4.9	µM

ADP binding			
$K_{d,ADP}$	0.116	± 0.034	µM

ATP binding*			
$K_{d,ATP}$	129.3*	± 25.8	µM

transient kinetics

mant-ADP binding			
$k_{on,mantADP}$	0.954	± 0.120	µM^{-1}s^{-1}
$k_{off,mantADP}$	0.1530	± 0.0004	s^{-1}
-> $K_{d,mantADP}$ (calculated ratio)	0.160	± 0.021	µM

mant-dADP binding			
$k_{on,mantdADP}$	0.817	± 0.043	µM^{-1}s^{-1}
$k_{off,mantdADP}$	0.0740	± 0.0001	s^{-1}
-> $K_{d,mantdADP}$ (calculated ratio)	0.091	± 0.005	µM

RNA dependent nucleotide binding

mant-dADP binding (+RNA)			
$K_{d,(RNA),mantdADP}$	0.095	± 0.019	µM

ATP binding (+RNA)*			
$K_{d,(RNA),ATP}$	5.04*	± 0.79	µM

RNA binding

RNA binding (-ATP)			
$K_{d,RNA}$	12.41	± 2.75	µM

RNA binding (+ATP)			
$K_{d,(ATP),RNA}$	0.136	± 0.020	µM

Of note, the global fit included the RNA titration- as well as all previous titration data. These previously determined binding affinities for (mant-) ADP/ATP were not included as fixed parameters, moreover the original primary titration data were used. In this approach these data were fitted numerically in contrast to the analytical fitting methods that were applied before (**see figure 3.3.4 – 3.3.7**). With exception of the described variance in $K_{d,(RNA),ATP}$, numerical and analytic fitting resulted in similar values for the kinetic parameters, showing robustness of the global fit (**see suppl. table 7.2**).

Enzymatic function of DDX1

3.3.8 ATPase activity of DDX1

DEAD-box helicases, like all other P-loop containing proteins, do hydrolyze ATP[200], though the energy of ATP hydrolysis is most probably not directly required for RNA remodeling and unwinding[73]. Although all DEAD-box helicases share the common RecA-like core fold, they show significant variation in hydrolysis rates between different orthologs[56].

Recombinant DDX1 exhibits an intrinsic ATP hydrolysis activity that was observed in a coupled ATPase assay[161] (**suppl. Figure 7.1.16**). This ATPase activity could be specifically attributed to DDX1 by measuring the apparent ATPase rate of the DDX1 K52A mutant. This K52A variant is abolished in its enzymatic activity, excluding that the observed activity of WT protein arose from potential contaminantions (**suppl. Figure 7.1.15**).

A minimal reaction scheme for ATP hydrolysis by DDX1 was established that includes a potential conformational change of the enzyme to an active conformation (**Figure 3.3.10**).

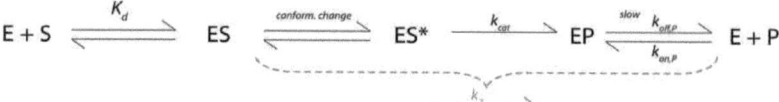

Figure 3.3.10 **Minimal reaction scheme for ATP hydrolysis** The enzyme (=E) DDX1 binds its substrate (=S) ATP, a process defined by the affinity constant K_d. The substrate bound complex (=ES) undergoes a potential conformational change to an activated conformation (=ES*). Then ATP is hydrolyzed (described by k_{cat}) and the product (=P) ADP is slowly released, due to the high ADP affinity of DDX1. In the classical Michaelis-Menten analysis all steps that lead from the enzyme-substrate complex (ES) to the enzyme product released state (E+P) are represented by the apparent rate constant k_2 (depicted in grey). Classically, it is assumed that dissociation of EP is fast and can be ignored in the forward reaction. For simplicity the influence of RNA was not incorporated in the scheme.

The apparent rate of ATP hydrolysis at different ATP concentration was measured under conditions of excess ATP over protein. Initial reaction velocities of individual hydrolysis traces at different ATP concentrations were fitted to the hyperbolic Michaelis-Menten-equation (Equation 6, **Figure 3.3.11 a**). An apparent $K_{m,ATP}$ = 1.75±0.33 mM was obtained, a value that is similar to the reported K_m for DDX1[135]. Notably, the determined $K_{m,ATP}$ for ATP is ten times higher when compared to the $K_{d,ATP}$ for ATP, suggesting that binding is not rate limiting and substrate turnover and subsequent release have an essential influence on the reaction velocity. Thus, $K_{m,ATP}$ would not be a simple measure for the ATP affinity (defined by $k_{off,ATP} / k_{on,ATP} = K_{d,ATP}$), but be influenced by all further processes leading to product formation and release (k_2 in **Figure 3.3.10**, i.e. $(k_{off,ATP} + k_2)/k_{on,ATP} = K_{m,ATP}$). This is in good agreement with the fact that the reaction product is ADP, which is bound much more tightly than ATP.

The maximal apparent hydrolysis rate $k_{cat(ATP)}$ = 0.096±0.005 s^{-1} (at saturating ATP) that was determined is low, but still within the lower range of what has been reported as intrinsic, i.e. unstimulated activity for other DEAD-box proteins[201]. Strinkingly, this hydrolysis rate $k_{cat(ATP)}$ is almost equal to the off-rate $k_{off,mantADP}$ for mant-ADP and the predicted off-rate $k_{off,ADP}$ for ADP (as estimated from the values for mant-ADP, **see section 3.3.5**). This suggests that ADP product release ($k_{off,ADP}$) is the rate-limiting step and the actual $k_{cat(ATP)}$ for ATP hydrolysis might be higher (compare k_{cat} and $k_{off,P}$ in **Figure 3.3.10**), but is not detected in the coupled assay that only monitors free ADP.

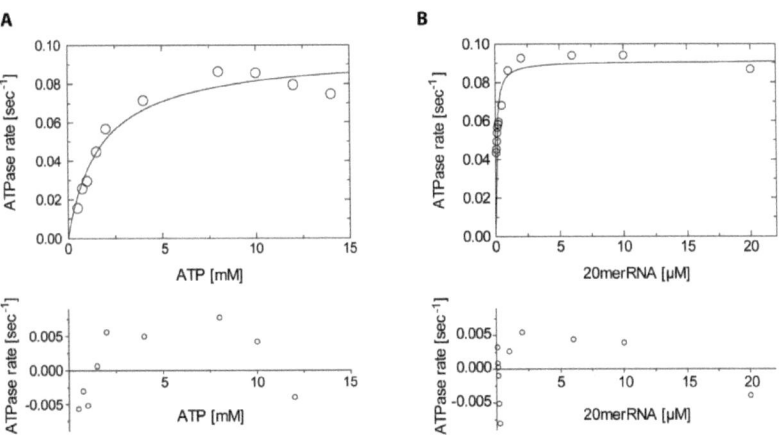

Figure 3.3.11 **ATPase activity of DDX1. A**, ATP concentration dependence of the intrinsic steady-state ATPase rate per molecule DDX1 is shown. For the assay 1 µM DDX1 was incubated with the respective concentrations of ATP-MgCl$_2$ and initial reaction velocities = maximal rates (per min) were plotted. These apparent rates were fitted to the hyperbolic form of the Michaelis-Menten equation. **B**, RNA concentration dependence of the ATPase rate, fitted with the quadratic equation (Equation 7). 1 µM DDX1 was incubated with indicated concentrations of a 20mer RNA in the presence of 2 mM ATP-MgCl$_2$. Residuals from the fits (i.e. Δy between fit and actual data) are shown beneath the plots, where the y-axis represents 10 % of the maximal reaction rate.

Different short ssRNAs were used to stimulate DDX1 ATPase activity. Stimulation of the hydrolysis rate did not differ for RNAs ranging from 10 to 20 nucleotides in size and which had a different sequence (**suppl. Figure 7.1.16**). When the stimulation of hydrolysis by tRNA was tested, a ten times higher concentration was required to achieve the same stimulatory effect as with unstructured ssRNA (**suppl. Figure 7.1.17**). Additionally, ssDNA did not influence ATPase activity (**suppl. Figure 7.1.16**).

A 20mer ssRNA (for sequence see materials and methods **section 2.3.9**) was used to determine the RNA concentration dependent stimulation of the ATPase rate under conditions of saturating (10 mM) ATP-MgCl$_2$ concentrations. Unexpectedly, a stimulation of the initial reaction velocity by RNA titration was not observed under those conditions. The maximal ATPase rate measured at 10 mM ATP-MgCl$_2$ in the presence of different RNA concentrations was the same as in the absence of the RNA (**suppl. Figure 7.1.18**) and Michaelis-Menten curves could not be measured. Therefore, RNA concentration dependent stimulation of ATPase activity was measured at 2 mM ATP-MgCl$_2$, which corresponds approximately to the $K_{m,ATP}$ of unstimulated DDX1. Under those conditions, ATP hydrolysis could unambigiously be measured in dependence of the concentration of the 20mer ssRNA with an apparent $K_{m,RNA}$ (at 2 mM ATP-MgCl$_2$) = 87.2±11.9 nM (**Figure 3.3.11 b**, Table 3.5). The maximal apparent hydrolysis rate obtained by RNA stimulation, $k_{cat(RNA)}$ = 0.11±0.0028 s^{-1} (at saturating RNA), is highly similar to $k_{cat(ATP)}$, the maximal apparent hydrolysis rate at ATP saturation in the absence of RNA. A possible explanation for this finding is that RNA promotes ATPase activity solely by stimulating ATP binding, but not hydrolysis. Another possibility is that the reaction is limited by ADP release, as suggested before, and a higher $k_{cat(RNA)}$ can not be detected, since only released ADP is monitored in the assay.

Under conditions of saturating RNA concentrations (40 µM), the maximal apparent hydrolysis rate was observed at all ATP concentrations within the concentration range of the experiments (0.5 – 10 mM ATP-MgCl$_2$, **suppl. Figure 7.1.19**). This finding is in line with the significantly increased ATP affinity in the DDX1-RNA complex. (see section 3.3.6) Thus, it seems that ATP and RNA mediated effects on ATP hydrolysis by DDX1 are tightly coupled.

Table 3.5. ATP hydrolysis activity of DDX1
Parantheses indicate the respective molecule that is present under saturating conditions. The errors represent the standard error from the fits. * at 2 mM ATP-MgCl$_2$

ATP titration			
$K_{m,ATP}$	1.75	± 0.33	mM
$k_{cat(ATP)}$	0.096	± 0.005	s^{-1}

RNA titration *			
$K_{m,RNA}$	87.2	± 11.9	nM
$k_{cat(RNA)}$	0.110	± 0.003	s^{-1}

3.3.9 Helicase activity of DDX1

DDX1 harbors the signature motifs of an RNA-helicase (**Figure 3.1.1**) and to test for a potential helicase activity, an unwinding assay was established. Initially up to 41 nucleotides long DNA/RNA-hybrids were used as substrate that have been used in studies of the helicase activity of DDX1 before[111]. Fluorescently (either FAM or Fl) labeled DNA oligonucleotides were annealed to RNA oligonucleotides to generate hybrid substrates. These hybrid substrates were incubated with DDX1 in the presence of nucleotides. Separation of the fluorescently-labeled from the unlabeled strand by a putative DDX1 helicase activity was analyzed by electrophoresis on native acryl-amide gels. To avoid immediate reannealing of separated strands, an excess of unlabeled competitor DNA was included in the reaction (**Figure 3.3.12 a**). Double-stranded and single-stranded substrate could clearly be separated on the gel. Incubation of the double-stranded substrate with DDX1 either in the absence of nucleotide or in the presence of ATP or ADP did not result in species, corresponding to single-stranded substrate (**Figure 3.3.12 a**). Instead, a shift of the fluorescencently labeled species to the gel-pockets could be observed upon incubation with DDX1. Staining of the acryl-amide gel with InstantBlue® showed that the protein is retained in the gel-pockets (**suppl. Figure 7.1.20**). This means that a fraction of the DNA/RNA substrate is complexed with DDX1, therefore trapped in the gel-pockets and does not migrate into the gel. Please note that buffer and running conditions were optimized for a proper separation of ssRNA and dsRNA in the experiment and that these conditions are incompatible with the EMSA conditions (**see section 3.3**) that would have allowed the protein to properly migrate into the gel.

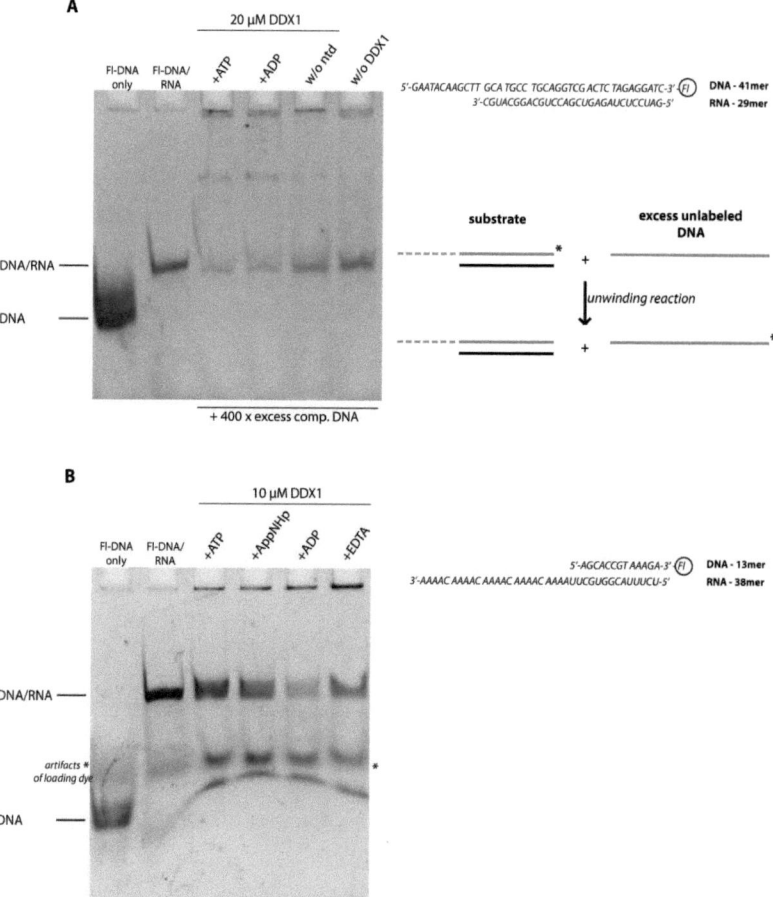

Figure 3.3.12 **DDX1 helicase displacement assay. A**, a solution of 20 µM DDX1-728 was incubated with 10 mM nucleotides and 0.25 µM DNA/RNA substrate as indicated. Reactions were incubated at 310 K for 60 min and then separated on a 15 % TBE-gel. To avoid re-annealing of unwound substrate a 400 fold excess of unlabeled competitor DNA was included in the reactions. DNA/RNA substrate was designed according to Li, Monckton and colleagues 2008[111] **B**, a solution of 10 µM DDX1-728 was incubated with 10 mM nucleotides and 0.25 µM DNA/RNA substrate as indicated. Reaction conditions were as in A. Black smear marked with * is an artifact of the dye used for gel-loading. DNA/RNA substrate was designed according to Jankowsky and colleagues, 2008[157].

To exclude that DNA/RNA substrate unwinding deficiency of DDX1 is due to long stretches of paired nucleotides and processivity issues, a substrate with a shorter duplex region and a single-stranded RNA overhang was designed, according to Jankowsky and colleagues 2008[157]. This substrate was incubated with DDX1 and different nucleotides (**Figure 3.3.12 b**). No unwinding activity could be observed.

Similarly to experiments with long substrates, the labeled species was retained in the gel-pockets, together with the protein (**suppl. Figure 7.1.20**).

To complement the helicase studies, in addition to the DNA/RNA hybrids, a dsRNA was used for the unwinding assay (**Figure 3.3.13**).

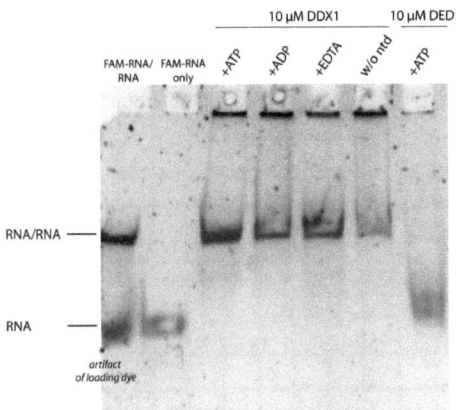

Figure 3.3.13 **DDX1 helicase assay with dsRNA substrate.** A solution of 10 µM DDX1-728 was incubated with 10 mM nucleotides and 0.5 µM dsRNA substrate as indicated. DED1, a DEAD-box protein from yeast that is involved in translation initiation and is supposedly active as a RNA helicase on short substrates, was included as a positive control[59] (suppl. Figure 7.3.12 (DED1 purification) and suppl. Figure 7.1.21 (DED1 helicase activity). For a negative control with the DED1 protein, see suppl. figure 7.1.20. Reactions were incubated at 310 K for 60 min and then separated on a 15 % (w/v) TBE-gel (pH 8.3). * marks an artifact of the dye used for gel-loading. The dsRNA substrate was designed according to Jankowsky and colleagues, 2008[157].

DDX1 was unable to unwind dsRNA substrates, but did bind them, evident through the fluorescence signal of a dsRNA/protein complex that was retained in the gel-pockets. In contrast, when the dsRNA was incubated with the yeast DEAD-box protein DED1 no substrate shift to the gel-pocket was observed. DED1 did unwind the duplex substrate (**Figure 3.3.13**).

In conclusion, the results from the unwinding assays show that recombinant DDX1 is unable to unwind short RNA/RNA and RNA/DNA duplexes under the buffer and reaction conditions used. In contrast, the homologous yeast DEAD-box protein DED1 was active under the given conditions. However, a general functional deficiency of recombinant DDX1 is unlikely as the protein was proven to be activity as an ATPase (**Figure 3.3.11**) and due to the cooperative binding of RNA and ATP (**see section 3.3.6**).

3.3.10 Summary of the functional studies

The results of the functional characterization of DDX1 revealed its unique enzymatic properties. DDX1 did exhibit an ATP affinity in the range of other DEAD-box proteins, but showed one of the highest ADP affinities to be ever reported. The affinity for ADP was higher by a factor of more than 1000 when compared to the affinity for ATP, but ATP affinity was stimulated by RNA binding. However, even in the RNA bound form the affinity of DDX1 for ATP was 50 times lower than the affinity for ADP.

Data sets from different spectroscopic titration experiments were combined in a global fit to access the RNA affinites of DDX1. In accordance with cooperative binding of RNA and ATP, an increased RNA affinity was obtained for the ATP bound form of DDX1. DDX1 was enzymatically active as an RNA stimulated ATPase, though, it showed a deficiency in its potential DEAD-box specific helicase activity. RNA stimulated ATPase activity further substantiated cooperativity in RNA and ATP binding.

4. Discussion

In this thesis the novel human DEAD-box protein DDX1 was characterized both structurally and functionally. This discussion is therefore splitted in three major parts. It starts with the structural characterization of the SPRY domain and its role in the context of the entire DDX1 protein. Then the functional characterization of DDX1 is discussed. Finally a conclusion is drawn from the presented results and an outlook on future developments is given.

Discussion

4.1 Structural studies

The SPRY domain structure – a modular interaction platform within DDX1

This work reports the structure of the SPRY domain of DDX1. The SPRY domain is a unique feature of the otherwise canonical DEAD-box helicase DDX1. The SPRY domain adopts a compact β-sandwich conformation and constitutes a modular interaction domain. In this section, the putative interaction surface and the role of the domain for the function of DDX1 will be discussed. Two putative scenarios for substrate binding to the SPRY domain are conceivable. It could directly mediate the interactions of DDX1 with other proteins or it could also enhance the RNA binding surface of DDX1.

4.1.1 Overall structure of the DDX1 SPRY domain reveals differences to other SPRY domains

The structure of the SPRY domain of DDX1 revealed the conventional SPRY fold[98], consisting of two layers of concave shaped β-sheets that stack together (**Figure 3.2.3**). Previous structural studies on SPRY domains have identified loops, which connect both β-sheets on one side of the domain as the canonical interaction surface A[101, 105]. In close proximity to these canonical interaction loops, in the SPRY domain of DDX1 a patch on the β-strand surface was identified that is positively charged and highly conserved. This surface region was therefore entitled extended surface A (**Figure 3.2.9**). Considering the overall structure of the SPRY domain of DDX1, the closest structural homolog is the SPRY domain from Ash2L[107], despite bearing low sequence similarity. Ash2L plays a critical role in the regulation of histone methyltransferases by interacting with components of the MLL protein family[202]. As the function of Ash2L in activating histone methylation is distinct from the cellular activities known for DDX1[110-112, 135], one can not deduce similarities in the proteins that interact with the respective SPRY domain. Indeed, whereas the core structure is almost identical, the loops that form interaction surface A show significant differences in their conformation and their length between the two proteins (**Figure 3.2.6**). Similarly, comparison of DDX1 SPRY with SPRY domains of SOCS box proteins (SPSB) in complex with interacting peptides[101, 105] revealed highest divergence in the peptide binding loops of surface A. Although a peptide interaction region within the DDX1 SPRY domain has not been determined, the binding mode must be different, if the conventional peptide binding region is used at all. Taking a closer look at two SPRY domain structures for which protein-protein interations have been reported in

Discussion

the literature[107, 203], further suggests that a putative interaction region within the DDX1 SPRY domain might differ and may not include the canonical interaction loops after all.

The first example is the Ash2L-SPRY domain, where five residues have been identified via site-directed-mutagenesis that are essential for the interaction with the protein RbBP5[107]. Based on this finding, the region encompassing the five residues has been suggested to be an RbBP5 binding pocket. This binding pocket, however, is not found in DDX1-SPRY.

The second example is the SPRY domain of the protein Trim5α. A number of structures of more distantly related SPRY domains of Trim (Tripartite motif) proteins have been reported[203-208]. Trim proteins regulate the innate immune response to infection and the SPRY domain of Trim5α restricts HIV-1 infection by binding to the retrovirus capsid[203, 207]. Comparison of the structure of the SPRY domain of Trim5α (PDB entry 2LM3[203]) with DDX1 SPRY (Z-Score of 13.2, overall rmsd of 2.3 Å for the alignment of 140 residues[192]) revealed largest structural differences in the loop between β3 and β4 (**Figure 4.1.1**).

Discussion

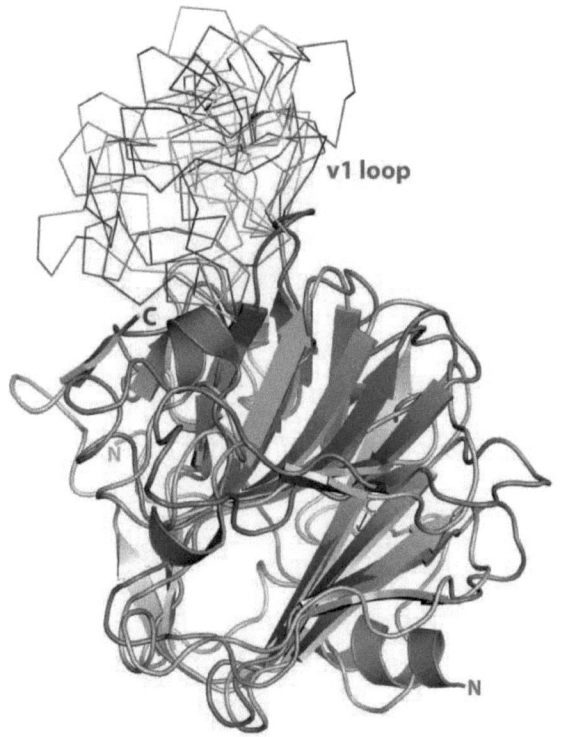

Figure 4.1.1 **Comparison of DDX1 SPRY with the distantly related SPRY domain of Trim5α** The superposition of DDX1 SPRY with Trim5α SPRY[203] as obtained from the DALI server[192] is shown. Both proteins are depicted in ribbon representation. DDX1 SPRY is depicted in green and Trim5α SPRY is depicted in blue. The region of highest divergence, the loop connecting β-strands 3 and 4 (referring to DDX1) is indicated by more intense color shading and labeled v1 loop. For the NMR structures of Trim5α SPRY loop v1, the different states i.e. models in the ensemble are indicated by differentially colored ribbon chains. The ten lowest energy conformations calculated using NOE restraints[203] are shown.

Whereas in the DDX1 SPRY structure this loop could be modeled unambiguously into the electron density, in the SPRY domain of Trim5α the loop, designated v1 (for variable loop 1), was found to be highly flexible and was not be resolved in the crystal structure[203, 207]. In the DDX1 SPRY structure the equivalent loop contacts the solvent and has low B factors, excluding that its rigid conformation is a crystallization artifact. In Trim5α several different conformations of loop v1 (which in Trim5α comprises 30 residues) could be determined via NMR measurements [203]. The different conformations of the mobile v1 loop confer structural plasticity, which may allow interaction with multiple epitopes[203, 209]. In contrast to all Trim SPRY structures, such a modular v1 loop – like structural adaptation to interaction partners is unlikely for DDX1 SPRY, since the corresponding v1 loop is short (five residues) and rigid

(**Figure 4.1.1**). Of note, similar to DDX1, in Trim5α an anti HIV-1 function (distinct from capsid binding) could be mapped to the SPRY domain, however, in contrast to DDX1[136] this HIV-1 restrictive function is unknown[210].

4.1.2 The SPRY domain might influence the enzymatic activity of DDX1 due to its central position within the helicase core

To enable a glimpse on the DDX1 protein in its entirety, the SPRY domain crystal structure was placed in a homology model of the helicase core. In other P-loop containing proteins, additional domains can regulate enzymatic activity by interaction with P-loop motifs[211, 212]. Such regulation has specifically been reported for kinases[213, 214] and GTPases[211, 215]. Furthermore, a lid subdomain covers the ATP binding site in the kinase group of P-loop containing proteins, which is not present in the RecA group[212]. In the conformation of the homology model, the SPRY domain would neither show a kinase or GTPase like interaction with the P-loop, nor would it form a lid (as in the kinase group) and sterically block ATP or RNA substrate binding (**Figure 3.2.11**). Nevertheless, the SPRY domain could still influence enzyme catalysis, since it is directly connected to the P-loop (containing the essential catalytic Walker A motif) with its N-terminal linker and to the strand that precedes motif Ia (involved in RNA binding) with its C-terminal linker, suggesting that it may be involved in communication between the ATPase site and the RNA binding surface. It has been speculated that such a signaling task is performed by the the conserved helicase motif III in other DEAD-box proteins[30, 216, 217]. However, the exact functional contribution of motif III is a matter of debate in the field as mutations in the conserved motif III sequence showed different effects in various DEAD-box proteins[30, 216, 218, 219]. In DDX1, the SPRY domain could support motif III or even take over its function. The direct linkage of the essential motif for ATP hydrolysis with an RNA binding motif in DDX1 would facilitate the transduction of potential allosteric signals[220, 221] via the SPRY domain between the catalytically important residues (**Figure 4.1.2**).

Discussion

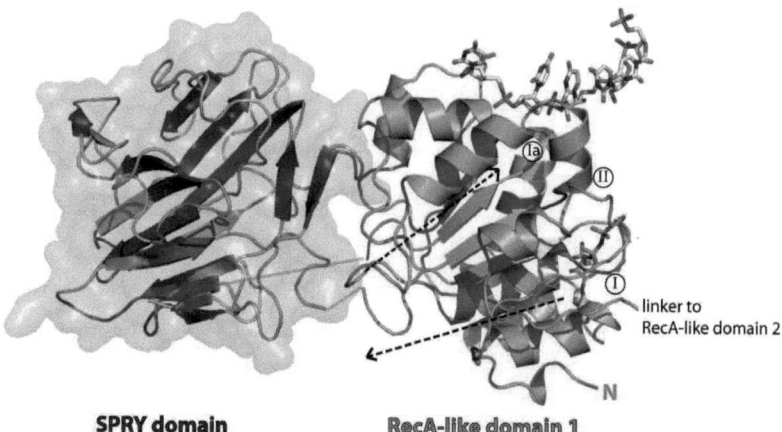

SPRY domain   **RecA-like domain 1**

Figure 4.1.2 **Putative communication pathways between conserved helicase motifs and the SPRY domain.** The DDX1 homology model is depicted in the same orientation as in figure 3.2.11, but a zoom-in on RecA-like domain 1 and the SPRY domain is shown. Conserved motifs I and II (Walker A and B) essential in ATP binding and hydrolysis are colored in red, motif Ia involved in RNA binding is colored in yellow. Residues that directly connect motif I and motif Ia to the SPRY domain are depicted in grey. Residues of interaction surface A are depicted in dark blue, residues of extended surface A in light blue. Red lines indicate the linkage of the SPRY domain to RecA-like domain 1. Black arrows indicated the course of the peptide strand from N- to C-terminus. RecA-like domain 1 is colored in green, the SPRY domain in brown. ATP and RNA are depicted as sticks and were placed in the model according to their positioning in the structure of VASA (PDB entry 2DB3) [31].

Facilitated by a potentially flexible linkage to the rest of the protein, the SPRY domain may alter its orientation upon transition between the different helicase states (**suppl. Figure 7.4.1**). For instance, in the "closed"-state of DDX1 the SPRY domain could enlarge the RNA binding surface on top of the two RecA-like domains. It may act as an extension of RecA-like domain 1 and support motifs Ia, Ib and Ic in RNA binding. A similar function has been observed for the C-terminal extension of yeast Mss116p, but in this DEAD-box protein the RNA binding surface of RecA-like domain 2 is extended[35]. The potential function of the SPRY domain in RNA binding is discussed in detail in section 4.1.5.

Extended insertion domains can also be found in other RNA helicases. In the structures of S. cerevisiae Mtr4 and Ski2 a large helical bundle is present at the C-terminus that is connected to the helicase core via a stalk containing β-barrel[222, 223]. These insertions constitute RNA binding domains, but are dispensable for the function of the helicase[222]. In contrast, the SPRY insertion domain in DDX1 is not protruding at the N- or C-terminus, but is positioned directly within the helicase core (**Figure 3.2.11 c**). It is therefore plausible that the SPRY domain insertion directly influences the helicase activity of the enzymatic core. Construct DDX1ΔSPRY, lacking the SPRY domain, displayed functional deficiencies in ATP hydrolysis, but it is unclear so far, whether they were caused by aggregation of the recombinant protein or do have a functional significance (**see appendix, section 7.3.6**).

Discussion

The homology model shown here (**Figure 3.2.11**), represents the "open"-state of DDX1, where the two RecA-like domains are flexible in their relative orientation to each other and are separated in space[224]. In this conformation, surface A and the region entitled extended surface A would both be freely accessible, suggesting that they could bind to putative interaction partners and thereby bring them in proximity to RecA-like domain 1 (**Figure 3.2.11**). The interaction of SPRY domains with other proteins will be discussed in the following two sections.

4.1.3 SPRY domains interact with DEAD-box helicases

Interestingly, reports are accumulating that show an interaction of SPRY domain containing proteins specifically with DEAD-box helicases[100, 225-227]. The best characterized example is the SPRY domain of *D. melanogaster* GUSTAVUS that interacts tightly with the DEAD-box RNA helicase VASA[100] (**Figure 3.2.6**). The SPRY-DEAD-box interaction is intriguing in comparison to DDX1, which is a DEAD-box helicase with a "built-in" SPRY domain. As the VASA 20-residue-interaction stretch is not conserved in DDX1, the helicase is most probably not able to bind to its own SPRY domain via surface A (**Figure 3.2.9**). In VASA the SPRY interaction sequence (comprising residues 184-203) is located in an N-terminal extension outside of the DEAD-box helicase core[44, 100]. As the structure of the VASA helicase only contains residues 202-621[31], there is no structural information of the SPRY interaction region in the context of the entire DEAD-box protein. Recent studies with the sea urchin *Strongylocentrotus purpuratus* GUSTAVUS protein suggested that it has two binding sites on *S. purpuratus* VASA and in addition to the binding site at the N-terminal extension, it also interacts with the DEAD-box core[228]. Mutational analysis indicated that *S. purpuratus* GUSTAVUS may interact with the N-terminal extension and the DEAD-box core of VASA independently through two separate binding surfaces[228].

Other recently identified SPRY-mediated protein-protein interactions include the helicases DDX41[225] and RIG-I [226, 227]. The SPRY domain of the E3 ubiquitin ligase TRIM21 interacts directly with RecA-like domain 1 of DDX41[225]. Only the core elements of RecA-like domain 1 of DDX41 are necessary for this interaction and extensions are dispensable[225].

The SPRY domain of the protein TRIM25, implicated in innate immunity signalling, interacts with the CARD-1 domain of the SF2 helicase RIG-I[206, 226, 227].

Discussion

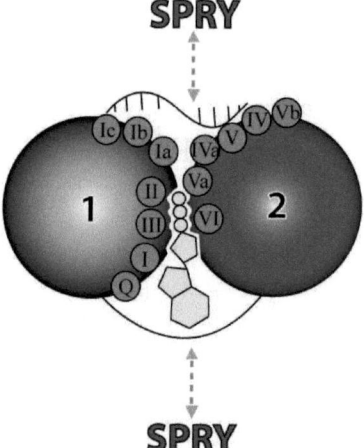

Figure 4.1.3 **Scheme for the interaction of an SPRY domain with the DEAD-box helicase core.** For several helicases a protein-protein interaction with SPRY-domain containing proteins has been shown. In these cases the SPRY domain (brown) interacts with the helicase core. DDX1 is a DEAD-box protein with a 'built-in' SPRY domain that could potentially interact with RecA-like domain 1 and thereby regulate enzymatic activity. RecA-like domains 1 (green) and 2 (blue) are drawn as spheres, connected by a short linker. Conserved helicase motifs, involved in ATP hydrolysis and in RNA binding are highlighted in orange. Note that the SPRY domain of DDX1 is connected to RecA-like domain 1 via a flexible linker.

The implications of the interaction of SPRY domains with helicases have not been investigated so far and interaction sites have not been mapped in detail. The only exception is *D. melanogaster* VASA, where a 20-residue peptide, corresponding to the motif at the N-terminal extension that interacts with the SPRY domain has been structurally characterized[101] (**Figure 3.2.6 b** and **d, Figure 3.2.7 b**). However, the second interaction motif within the helicase core has not been identified yet. Also in other DEAD-box proteins, the region in the helicase core, responsible for the protein-protein interaction with SPRY domains has remained elusive.

In DDX1 two scenarios for the role of the SPRY domain with respect to its potential interaction with the helicase core are conceivable. Firstly, DDX1 might have acquired the SPRY domain from one of its interaction partners by domain swapping during evolution. A domain swap could rescue a specific function that was lost by mutation[229] or regulate protein folding and aggregation[230]. Secondly, it could also be possible that the SPRY domain insertion serves a helicase regulating function. Instead of recruiting protein factors, it may interact with the DDX1 helicase core that is in immediate proximity on the same polypeptide chain (**Figure 3.1.1**). This may constitute an auto-regulatory mechanism (**Figure 4.1.3**). Future experiments will be directed on the exact influence of this domain on DDX1 activity. If it is possible to extract details from the mechanism of the intramolecular "SPRY domain – DDX1 helicase core" interaction and regulation, this would greatly advance the understanding of other DEAD-box

Discussion

proteins. Since DDX1 represents a unified SPRY-helicase in one polypeptide chain, it is an ideal model for studying the so far unknown relationship between SPRY domains and DEAD-box helicases. A detailed description of the helicase core - SPRY interaction region has to wait until structural information on full-length DDX1 are available, but details on the protein-protein interaction site of the DDX1 SPRY domain are discussed in the following.

4.1.4 A conserved SPRY surface patch as a potential protein-protein interaction region

In addition to their specific interaction with DEAD-box proteins in some cases, SPRY domains have generally been described as protein-protein interaction platforms[98, 99]. It is plausible that the SPRY domain within DDX1 also mediates interactions of the helicase with other proteins. Specifically the fact that DDX1 is found in the multi-protein HSPC117 complex[108, 114] implies that protein-protein interactions have to be formed, which may involve the SPRY domain.

In DEAD-box helicases the relative orientation of the two RecA-like domains is crucial for the activity[31, 36], but is modulated by protein-protein interactions of the helicase core. Prominent examples include eIF4AIII in the exon junction complex (EJC)[23, 24] and Dbp5 in the nuclear pore complex (NPC)[25, 231] that both alter their conformation upon binding of interaction partners to the helicase core. In contrast to all other DEAD-box proteins, DDX1 harbors the additional SPRY domain. This domain may establish protein-protein interactions independent of the two RecA-like domains of the helicase core. Such a mechanism would allow the RecA-like domains that are not involved in the interaction, to freely adopt different conformations and for the helicase core to perform its catalytic function unaffected by the interactions. This furthermore suggests, that inclusion of the SPRY interaction module might have equipped DDX1 with the possibility to interact with other proteins irrespectively of the conformational state of the helicase core ("open"- or "closed"-state), since it would simply not be involved in the SPRY domain – protein interactions. Supporting the role of the DDX1-SPRY domain in mediating protein-protein interactions, in studies, where specific fragments of DDX1 have been shown to interact with other proteins[93, 109, 110] (**also see suppl. Figure 7.6.1**), those peptide fragments always comprised the SPRY domain.

Structural comparison of the DDX1-SPRY domain with other SPRY domain structures has indicated differences in the canonical protein-protein interaction loops (**see section 4.1.1**). At a closer inspection of the sequence of the interaction loops, a difference in a putative DDX1-SPRY interaction surface becomes even more apparent. Notably, binding studies on other SPRY domain containing proteins have shown that protein-protein interaction sites are highly conserved to the extent that proteins are functionally interchangeable throughout mammalian species [105, 205]. In the SPRY domain of SPSB proteins, this conservation of protein-protein interaction sites leads to the recognition of a

Discussion

"ELNNNL/DINNNN" consensus motif in the respective interaction partner [101, 106]. Therefore, one could assume that the residues that form a protein-protein interaction site in the DDX1 SPRY domain are also highly conserved. However, at closer inspection of the sequence conservation two arguments indicate that the canonical interaction surface A does not exist in the DDX1 SPRY domain. Firstly, its sequence is not conserved from other SPRY domains to DDX1 SPRY (**Figure 3.2.7**). Secondly, the residues that would correspond to surface A in DDX1 SPRY are not conserved amongst DDX1 homologes from different model organisms (**Figure 3.2.8**). This suggests that surface A is not the protein-protein interaction region of the SPRY domain of DDX1.

Moreover, high conservation amongst DDX1 homologs from different model organisms was detected in the extended surface A region (**Figure 3.2.8**). In light of the pronounced conservation of the protein-protein interaction region across species in other SPRY domains, one might therefore assume that extended surface A mediates protein-protein interactions. Extended surface A is not conserved among SPSB SPRY domains and is an exclusive feature of DDX1 (**Figure 3.2.7**). The extended surface A might be a novel feature of the DDX1-SPRY domain and replace the canonical surface A as interaction region.

In general, protein-protein interaction interfaces are either characterized by hydrophobic residues, particularly aromatic residues, or they are mediated by opposite charge pairs[232-234]. Extended surface A contains only two surface exposed hydrophobic residues, one of them (W119) being aromatic, but instead harbors positively charged residues leading to an overall highly positive surface charge (**Figure 3.2.9**). Such a conserved, positive electrostatic surface potential has also been observed for the peptide binding site of SPRY domains from mammalian SPSB proteins[105], further strengthening a similar role of DDX1's extended surface A in mediating protein-protein interactions.

Although SPRY domains have been described as protein-protein interaction platforms[98] and the SPRY domain within DDX1 therefore may well serve such a function, insertion domains in DEAD-box helicases usually serve RNA-binding functions[35, 42, 185]. Such a role for the DDX1 SPRY domain can not be excluded and will be discussed in the following section.

Discussion

4.1.5 The possible role of the SPRY domain in tethering RNA substrates to the DDX1 helicase core

In many DEAD-box proteins, additional domains were shown to function in tethering RNA substrates[40-42] or in extending the RNA binding site of the helicase core[35, 39]. This suggests that instead of providing a protein-protein interaction platform within DDX1, the SPRY domain may also serve as an RNA binding platform. Whereas DNA binding sites can be identified by their specific fold that is able to interact with the major groove of dsDNA[235-237], for RNA that predominantly exists as a single stranded species and is therefore structurally more flexible, binding sites on the protein are much more difficult to predict [238-240]. In contrast to DNA interactions that mostly involve the phosphate backbone, RNA interactions are more often established via specific interaction with the bases or specific secondary structures (independent from the sequence) are recognized[240, 241].

For the human DEAD-box protein DDX3X, which is the only other DEAD-box protein that harbors an insertion at an equivalent position than the SPRY domain in DDX1 within RecA-like domain 1, a function of the insertion domain in RNA binding has been discussed[185]. In the structure of DDX3X the insertion creates an elongated, positively charged cavity that is positioned close to the 3'-end of a potential bound RNA substrate and is therefore expected to be involved in RNA binding and positioning[185]. The SPRY domain of DDX1 may likewise be involved in positioning of RNA substrate, possibly via the conserved, positively charged extended surface A. For human DDX19 a patch of highly positive surface potential was suggested to mediate the interaction with mRNA[48].

In general, in the "open"-state of DEAD-box proteins, duplex RNA substrate is exclusively bound by RecA-like domain 2[36]. In contrast, for DDX1 Liu and colleagues reported that a construct, lacking RecA-like domain 2 binds to viral dsRNA[137]. It is therefore tempting to speculate that in this construct the canonical RecA-like domain 1 does not bind RNA, but the SPRY domain insertion mediates the RNA binding.

For the DEAD-box proteins YxiN and HerA additional domains supplementing the helicase core have been shown to adopt an RRM fold[40, 242-245] (**also see Figure 1.5**). RRM are one of the few domains, where the fold is a strong hint for a function in RNA binding[238]. The SPRY domain of DDX1 does not adopt an RRM fold, but nevertheless it is interesting to note that additional domains of DEAD-box proteins can clearly function in RNA binding. Structures of these DEAD-box supplementing RRM domains in complex with short RNA substrates have revealed base specific interactions with few signature nucleotides, whereas the majority of interactions is unspecific via electrostatic interactions with the phosphate backbone[39, 42]. In YxiN lysine residues of the RRM generate an electropositive stretch that establishes salt bridges to the RNA phosphate backbone[39]. Despite not adopting the RRM fold, the SPRY domain of

DDX1 similarly harbors several lysine residues in extended surface A (**see Figure 3.2.6**) that also generate a positive surface patch (**see Figure 3.2.8**).

Even though from the current opinion on SPRY domains, being protein-protein interaction modules is more likely, future studies, however, will show, whether a role in RNA binding can be excluded – up to now RNA tethering may be possible. Independent of its role in protein- or RNA binding, the linkage of the SPRY domain to the remainder of DDX1 is of particular interest and will be highlighted in the next section.

4.1.6 N- and C-termini of the SPRY domain may constitute a flexible linker

SPRY domains are found as modular insertions in a number of different proteins[204, 205, 208, 246-250]. Structural data is available only on isolated SPRY domains so far and they have never been crystallized in the context of the full polypeptide chain of their associated proteins. Thus, information on the linkage of the SPRY domain to the rest of the protein is lacking[98]. Inability to crystallize proteins containing a SPRY domain may already hint at its flexible linkage to the rest of the polypeptide chain.

In DDX1, the regions that connect RecA-like domain 1 to the SPRY domain (residues 69 to 84) and the SPRY domain to RecA-like domain 1 (residues 270 to 285) show low conservation (**Figure 3.1.1** and **Figure 3.2.8**). These regions are not part of the compact SPRY domain fold and they can also not be assigned to the helicase core, suggesting that they correspond to linkers that connect the SPRY insertion with DDX1. In partial proteolysis experiments protease cleavage was mainly detected in these regions (**Figure 3.1.4**), indicating that they are flexible. Furthermore, at both, N- and C-terminus, residues corresponding to the potential flexible linkers were found to be disordered in the crystal structure and could not be modeled unambiguously into the electron density (**see section 3.2.3**). For the N-terminal linker region secondary structure prediction suggested an α-helical region (**Figure 3.1.1**), which would slightly restrict flexibility, but the respective residues in the structure do not form a helix (**Figure 3.2.3**). For the C-terminal linker region, the last five residues pack against strand β1, form strand β16 and enlarge β-sheet 1, whereas they can not be modelled in the other chain (**Figure 3.2.3 b**). However, by sequence alignment those C-terminal residues are not associated with the SPRY domain(**Figure 3.2.7**) and in other SPRY domains the C-terminus normally comes together with the N-terminus and point away from the SPRY β-sandwich (**Figure 3.2.6 c**). It is likely that the conformation of this β-addition module of C-terminal residues is an crystallization artifact, caused by removing the SPRY domain from the rest of DDX1 and that those residues are part of a linker region in the full-length protein.

Flexible linkers are found in many proteins, e.g. they form the hinge region in immunoglobulins and they are usually enriched in proline, serine and threonine residues[251]. In protein kinases, hinge regions, characterized by glycine repeats, allow domain movements that are important for the catalytic function[252-254]. The potential N- and C-terminal linker regions of the SPRY domain do not harbor long glycin repeats, but nevertheless contain many proline and glycin residues that would allow flexibility (**Figure 3.2.8**). Flexible linkage of the SPRY domain to the helicase core would enable domain movement[252, 255]. The SPRY domain may change its relative orientation to the rest of the protein and act as a gatekeeper by obstructing the active site or by interacting with residues essential for catalysis (**Figure 4.1.3**). Furthermore, the linker regions directly connect the important motifs of the helicase core with the SPRY domain and could therefore be important to transduce signals as discussed before (**see section 4.1.2**).

As mentioned above it is important to note that the potential SPRY linker regions are unconserved across different species (**Figure 3.2.6**). Residues involved in protein binding or fold stabilization are usually conserved[256] as they are exposed to evolutionary selective pressure, which would not be the case for the linker regions. It is not clear whether variations in the linker residues reflect differences in SPRY domain flexibility which may influence enzymatic parameters. Especially, the *C. elegans* DDX1 homolog shows profound differences in the linker regions, as they display sequence deviations and are considerably shorter when compared to the human ortholog. Irrespectively of individual sequence differences, flexible linkage of the SPRY domain to the remainder of the DDX1 protein might be an essential functional characteristic.

4.2 Functional studies

Distinctive mechanistic features of DDX1 - synergistic effects of ATP and RNA binding

The results of the detailed biochemical and biophysical characterization of DDX1 as described in section 3.3, revealed unique characteristics when compared to other DEAD-box helicases – an unusual high ADP affinity and a tight coupling between RNA- and ATP binding. In the following, a model is presented that can explain the cooperativity between RNA- and ATP binding to DDX1. Furthermore, the implications of the nucleotide binding and of the enzymatic activities in ATP hydrolysis and RNA-duplex-unwinding will be discussed.

4.2.1 ATP and RNA both have the propensity to induce the "closed"-state of the helicase

Based on the experimental data obtained from the functional analysis of DDX1 (reported in section 3.3) a model was developed to explain the observed cooperativity in ATP and RNA binding to DDX1 (**Figure 4.2.1**). In this model either RNA or ATP binding alone are sufficient to induce a conformational transition in DDX1 from the "open"- to the "closed"-state of the helicase. Formation of the "closed"-state would result in an increase of affinity for both substrates.

Discussion

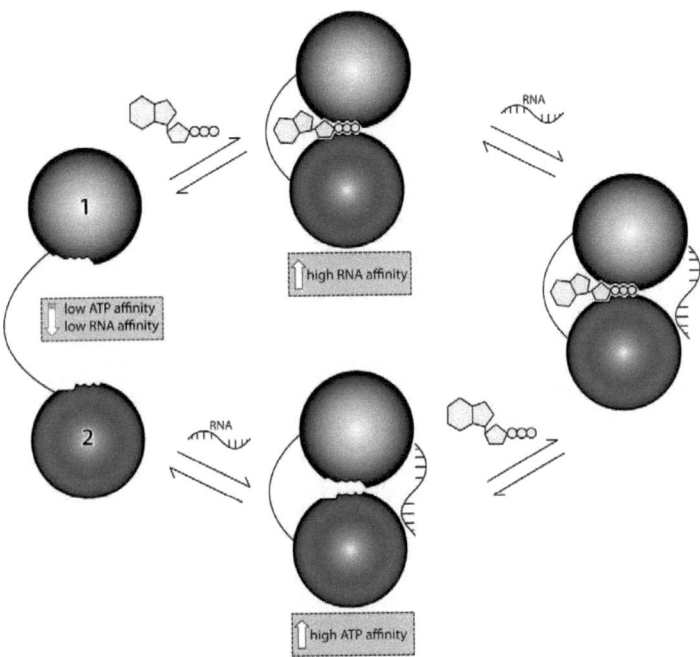

Figure 4.2.1 **Model for cooperative ATP and RNA binding to DDX1.** According to the model either ATP (upper pathway) or RNA (lower pathway) binding alone can induce the "closed"-conformation of the DDX1 helicase, thereby increasing DDX1's affinity for the respective other molecule. Numbers 1 and 2 delineate RecA-like domains 1 and 2. For simplicity the SPRY domain insertion in RecA-like domain 1 is omitted.

The model as presented in Figure 4.2.1 is based on experimental findings from both the ATPase activity assay and the equilibrium titrations and gives a framework to the observed effects.

The steady-state ATPase activity assay (**see results section 3.3.8**) showed strong coupling of RNA binding and ATP hydrolysis – whereas, at saturating RNA concentrations, the maximal hydrolysis rate was observed already at relatively low ATP concentrations, at saturating ATP concentrations no increase in the apparent ATPase rate by addition of RNA could be detected.

Furthermore, the results from the equilibrium titration experiments (**see results section 3.3.6**) are also in accordance with cooperative binding of ATP and RNA to DDX1. They showed an increased ATP affinity for the RNA bound form of DDX1 and an increased RNA affinity in the ATP bound state.

In general, only in the "closed"-state of DEAD-box proteins, a functional ATPase site and the full binding surface for RNA are assembled[31, 36]. It is therefore conceivable that the observed stimulation of ATP hydrolysis by RNA binding of DDX1 is achieved via induction of the conformational transition from the "open"- to the "closed"-state upon RNA binding (**Figure 4.2.1** lower panel). In the "closed"-, RNA bound state, DDX1 is then hydrolysis competent. Furthermore, the "closed"-state has the highest ATP affinity. In this scenario the main factor for stimulation of ATP hydrolysis would be this increase in ATP affinity. High ATP affinity of an RNA bound "closed"-state is also supported by the results from the equilibrium titration experiments (**see section 3.3.6**).

Similarly, ATP binding could also induce the transition to the "closed"-state (**Figure 4.2.1** upper panel). Formation of the "closed"-state by ATP binding would explain the increased RNA affinity that is detected in the presence of ATP. As observed in X-ray structures of DEAD-box helicases, in the "closed"-state RNA is bound by both RecA-like helicase domains, forming a high-affinity RNA binding site[23]. In contrast in the "open"-conformation of the ADP- or apo-state, only RecA-like domain 2 binds the RNA[36]. Such a mechanism of conformational transition induced by ATP binding is in full agreement with results from the ATPase assay (**Figure 3.3.10 b**). At high, saturating ATP concentrations the "closed"-state is formed and no further stimulation of hydrolysis by RNA is observed.

Furthermore, global fitting analysis of all recorded equilibrium titration data showed that ADP binding does not increase the RNA binding affinity of DDX1, as also apparent from the minimal binding scheme ($K_{d,ADP} = K_{d,RNA,ADP}$, i.e. $K_{d,RNA} = K_{d,ADP,RNA}$ see **Figure 3.3.8**), since RNA did not influence ADP binding. This pronounced difference of observed RNA affinity in the ATP-bound and ADP-bound/nucleotide-free state further supports the model in which DDX1 cycles between a high RNA affinity ATP-bound "closed"-state and a lower RNA affinity ADP-bound "open"-state. In this case, the "closed"-state would represent the "active" form of the enzyme, whereas the lower RNA affinity ADP-bound or nucleotide free "open"-state would represent an "inactive" form.

RNA helicases can either function as processive remodelers or immobile clamps[18, 19, 37]. The observed mechanism of cooperative ATP and RNA binding to DDX1 rather supports a function of DDX1 in RNA clamping than in processive unwinding or remodeling (**see section 4.2.6**). Recently functions in substrate clamping have been reported for a number of DEAD-box proteins[19, 90]. In this scenario, ATP hydrolysis by DDX1 is required prior to release of the RNA substrate and enzyme recycling.

Importantly, it has to be mentioned, that the K_d values for ATP, calculated from the equilibrium titrations, are inherently flawed, because probably they do not represent a simple displacement of mant-ADP by ATP. This becomes apparent at closer inspection of the ATP hydrolysis. The measured $k_{cat(ATP)}$ which is almost equal to the $k_{off,mantADP}$ for mant-ADP and the expected $k_{off,ADP}$ for ADP suggests that ADP product release is the rate-limiting step of the ATP hydrolysis reaction. Therefore, under steady-state conditions DDX1 would be mostly in the ADP bound state. This would imply that in the equilibrium titration experiments addition of ATP leads to an accumulation of ADP bound enzyme, in spite of the presence of the ATP regeneration system, since ADP release would be slower than ATP hydrolysis by DDX1. The ATP equilibrium titration experiment would then not simply measure a $K_{d,ATP}$ for ATP binding, but a combination of ATP binding and the following steps. As a consequence the calculated K_d values for ATP have to be handled with care and are difficult to interpret. The actual K_d values for ATP might be smaller, meaning that the ATP affinity would get underestimated.

Nevertheless two things can be clearly stated from the observations. Firstly, there is a huge preference of DDX1 for ADP over ATP and a large difference in affinity between the nucleotides. Secondly, the presence of RNA leads to an ATP affinity increase irrespective of the exact K_d value. Thus, cooperativity in binding of ATP and RNA is definitely an essential characteristic of DDX1.

Discussion

4.2.2 Adenosine-nucleotide affinity of DDX1 – escape from an ADP stalled complex

The previous section focused on the relationship between ATP and RNA binding to DDX1, however, already at closer inspection of the different affinities for ATP and ADP interesting features become evident. For several DEAD-box helicases nucleotide affinities have been determined experimentally[60]. DDX1's binding affinity for ATP is within the range of what has been reported[57, 58], even when taking inaccuracies of the ATP equilibrium titrations (due to a high steady-state fraction of ADP bound enzyme) into account (**Table 4.1**). In contrast to ATP, DDX1's affinity for ADP is extremely tight and higher by a factor of almost 500 when compared to ADP affinities of other DEAD-box proteins[57] (**Table 4.1**). Strikingly, the difference in affinity between the two nucleotide species is in the range of three orders of magnitude. Such a pronounced difference in the affinities between ATP and ADP has not been observed for any other DEAD-box helicase to date[34, 45, 60, 196]. It can be assumed that the ADP bound form represents the inactive state of DDX1, since only in the ATP bound form a high affinity for the RNA substrate is established, allowing a productive working cycle (**see section 4.2.1**).

In the cellular environment, the ATP concentration is around 1-10 mM and hence it is five- to ten-fold higher when compared with the concentration of ADP[257-259]. Thus, the high ADP affinity of DDX1 most likely causes the protein to be predominantly found in the ADP bound form and the ATP-state to be barely populated. However, this would render DDX1 as an inactive protein (unable to undergo repeated cycles of ATP hydrolysis), and the existence of a nucleotide-exchange-factor (NEF) reminiscent of G-proteins for DDX1 is conceivable. Such an exchange factor has also been described for the yeast DEAD-box protein Dbp5 (human DDX19)[67], where the N-terminal domain of Nup159 (Nup214 in *H. sapiens*) binds to Dbp5, opens the RecA-like domains and leads to ADP release[25, 67] (**Figure 4.2.2**). A similar mechanism might apply for DDX1 as well and would facilitate to overcome inactivation, caused by the high ADP affinity. Such a mechanism with an external NEF could provide means to regulate DDX1[19, 260]. Similar as described for Dbp5[19, 67, 260], regulated ADP release may prevent off-target activity, i.e. binding to any RNA that is not a designated DDX1 target. This could be essential for spatial activation, since as this study shows, DDX1 is promiscuous in its RNA specificity (like most DEAD-box proteins[37]) and as previously shown, it is found in several cellular compartments, where it is involved in different cellular processes[91, 117].

Recently it has been demonstrated that NEFs not only facilitate spatial activation, but in the case of small GTPases are directly responsible for specific targeting of their associated enzyme to a designated cellular compartment[261]. In this example the GTPase specific NEF harbors a subcellular localization signal and recruits its associated GTPase to specific membrane compartments[261, 262]. Similarly DDX1 might be recruited specifically to its designated place of action – for instance one might imagine localization to the nucleus, where tRNA splicing takes place in human cells[263] – by an associated NEF. This would be of special importance, since DDX1 does not harbor a nuclear localization signal (NLS). For

Discussion

yeast, where tRNA processing involves other pathways and takes place in the cytoplasm[264, 265], there would be no need for a transport, however, no DDX1 homologue could be detected in S. cerevisiae anyways. Since DDX1 is found in a complex with other factors in the tRNA processing HSPC117 complex[108], one of those proteins could potentially function as such a NEF. NEFs can modulate the thermodynamic and kinetic properties of their associated enzymes, when incorporated in a ternary complex with respect to a binary complex consisting only of enzyme and substrate[266, 267]. A putative NEF for DDX1 should fulfill two functions, it should accelerate the ADP release and it should shift the affinity towards ATP.

In principle, in accordance with the cooperativity model, RNA may bind to the ADP-form of DDX1, thereby stimulating ATP affinity and inducing nucleotide exchange. However, it is questionable, whether the small amount of RNA that is bound in the low-RNA affinity ADP-form of DDX1 is capable of achieving this. Furthermore, even in the RNA bound form, there is still a pronounced preference for ADP (by a factor of 50) over ATP. This rather suggests that other factors than RNA like e.g. a NEF must come into play. Such factors would help to escape a "hen and egg" scenario: which binds first, RNA or ATP? RNA is preferentially bound in the ATP-form, ATP affinity is increased in the RNA-bound form of DDX1, but where to start?

4.2.3 Comparison of the cooperativity in ATP and RNA binding with other DEAD-box proteins

To enable a classification of the different nucleotide affinities and cooperativity effects, observed for DDX1, experimentally determined affinity constants were compared with those available in the literature (**Table 4.1**). DDX1 showed a significantly increased affinity for ATP in the presence of saturating amounts of RNA (**see results section 3.3.6**). Elevated ATP affinities have been described for RNA-bound-states of other DEAD-box proteins, though the changes in affinity have been reported to be modest [57, 58] (**Table 4.1**). One may anticipate that in the presence of RNA substrates, DDX1 requires ATP in order to exert its function - be it simple clamping, translocation or unwinding – thus, rationalizing the coupling of RNA binding to an increase in ATP affinity.

Interestingly for the human eIF4A (the traditional model DEAD-box helicase) a similar situation as with DDX1 was observed[75]. In both proteins, the RNA affinity is higher in the ATP bound state when compared to the nucleotide free or ADP bound state. However, in eIF4A the overall RNA affinity is much lower[75] (**Table 4.1**).

Discussion

Table 4.1 Comparison of nucleotide affinities of DEAD-box proteins. (RNA) means affinities were determined in the presence of RNA, (mant) means affinities have only been determined for mant derivatives.

DEAD-box protein	$K_{d(ATP)}$	$K_{d(ADP)}$	$K_{d(RNA)}$	$K_{M(ATP)}$	k_{cat}
DDX1 (H. sapiens)	129 µM	0.11 µM	12.4 µM	1750 µM	0.1 s^{-1}
DED1 (S. cerevisiae)	n.a.	n.a.	0.02 µM[59]	300 µM[59] (RNA)	n.a.
Mss116 (S. cerevisiae)	126 µM[58]	43 µM[58]	0.117 µM[58]	250 µM[58] (RNA)	0.26 s^{-1} [58]
DbpA (E. coli)	72 µM[57]	51 µM[57]	0.016 µM[57]	65 µM[57] (RNA)	0.01 s^{-1} [57, 58]
eIF4A (H. sapiens)	80 µM[75]	30 µM[75]	5 µM[75]	80 µM[75]	n.a.

Influence of ATP and RNA on binding affinities

DEAD-box protein	$K_{d(ATP)}$ +RNA	$K_{d(ADP)}$ +RNA	$K_{d(RNA)}$ +ATP	$K_{d(RNA)}$ +ADP	$K_{M(RNA)}$ +ATP	k_{cat} +RNA
DDX1 (H. sapiens)	5.0 µM	0.1 µM (mant)	0.136 µM	12.4 µM	0.087 µM	0.11 s^{-1}
DED1 (S. cerevisiae)	n.a.	n.a.	0.012 µM[59]	n.a.	n.a.	5.6 s^{-1}[59]
Mss116 (S. cerevisiae)	80 µM [58]	92 µM [58]	0.047 µM[58]	0.336 µM[58]	0.2 µM[58]	1.5 s^{-1} [58]
DbpA (E. coli)	51 µM[57]	23 µM[57]	0.011 µM[57]	0.01 µM[57]	0.001 µM[57]	5 s^{-1} [57]
eIF4A (H. sapiens)	10 µM[75]	8 µM[75]	1 µM[75]	40 µM[75]	1 µM[75]	0.017 s^{-1} [75]

In general, coupling of ATP and RNA binding has been reported for other DEAD-box proteins as well[78, 80], but the cooperative effects for DDX1 are the most pronounced that have been observed so far (**Table 4.1**) and a cooperativity model as introduced for DDX1 in this thesis (**Figure 4.2.1**) has not been proposed for any other DEAD-box protein before.

At closer inspection of the different DEAD-box proteins, depicted in table 4.1, one interesting example is the bacterial DEAD-box protein DbpA. For DbpA a situation is encountered that is very much in contrast to the cooperative mode of nucleotide binding in DDX1. DbpA is specifically activated in its ATPase activity by 23S rRNA (500-2000 fold increase in k_{cat})[57, 72, 268]. However, in this case, RNA binding does not significantly increase the ATP affinity[57] (**Table 4.1**), which implies that ATPase stimulation is achieved by different means than in DDX1 and here the cooperativity model does not apply.

Discussion

Other DEAD-box proteins that display cooperativity effects similar to DDX1 include e.g. DED1 and eIF4A (**Table 4.1**). For those two proteins conserved helicase motif IV has been reported to be important for the cooperativity between ATP and RNA binding[218, 269]. Motif IV can only contribute to cooperativity in the "closed"-state. This means the fact that cooperative effects mediated by Motif IV have been observed for DED1[59, 269], and eIF4A[75, 269] implies that either ATP- or RNA binding can induce domain closure. This linkage of cooperativity effects with formation of the "closed"-state of DEAD-box proteins is very much in line with the DDX1 cooperativity model (**Figure 4.2.1**).

Only in a few cases coupling between RNA and ADP binding has been reported for DEAD-box proteins[270], which is not observed in DDX1. This absence of coupling suggests that ADP does not have the propensity to induce structural changes in DDX1 as e.g. transition to a higher RNA affinity "closed"-state of the helicase. This is in agreement with studies on eIF4A, where the γ-phosphate group of ATP has been proposed to mediate changes in the protein conformation[194]. Similarly, in the RNA induced "closed"-state, preferred binding of ATP versus ADP may be mediated by additional contacts with the γ-phosphate group.

Studies with the *Bacillus subtilis* DEAD-box helicase YxiN have directly addressed cooperative binding of ATP and RNA by FRET experiments[80]. It has been reported that both, ATP- and RNA binding are required for formation of the "closed"-state of the helicase[77, 80]. Such a situation would be in contrast to the DDX1 cooperativity model and to the experimentally observed intrinsic ATPase activity of DDX1 in the absence of RNA substrate, since the "closed"-state has to be assembled in order to create the ATPase active site[36]. Conserved helicase motif VI in RecA-like domain 2 can only make its essential contributions to ATP hydrolysis if both helicase domains come in close proximity[34]. A construct of DDX1 that was devoid of motif VI at the N-terminus (by truncation) was ATPase deficient (**see appendix section 7.3.1**). Therefore, one may rather assume a functional difference between the bacterial YxiN protein and human DDX1 (that can presumably adopt the "closed"-, ATPase active state even in the absence of RNA). In line with this hypothesis YxiN does barely exhibit any intrinsic ATPase activity[271].

Structural studies on other helicases also support the DDX1 cooperativity model. X-ray structures of the SF1 helicase Upf1 in the AppNHp bound state have revealed that ATP binding alone can already induce the "closed"-state of RecA-like domains 1 and 2[38]. In contrast, the ADP bound state of Upf1 did not show domain closure. Similar observations have been made in the ADP bound structure of the DECD-box helicase UAP56, which adopts the "open"-state[26]. Domain closure in the ATP bound state without RNA being present, but maintenance of the "open"-state with ADP bound, as observed in these structures, is in line with the cooperativity model. Furthermore, maintenance of the "open"-state, could explain the lack of influence of ADP on RNA affinity in DDX1.

Discussion

In addition to the functional and structural studies on other DEAD-box helicases, some reports have also specifically investigated mechanistic details of DDX1. The following sections will focus on enzymatic activities of DDX1 and refer to what is known on the specific function of this protein in the literature so far.

4.2.4 RNA binding - Sequence specificity

The helicase core of DEAD-box proteins interacts exclusively with the sugar phosphate backbone of oligonucleotides[31], resulting in sequence unspecific RNA binding per se[13]. In DDX1 this helicase core is supplemented with the SPRY domain (**Figure 3.1.1**) and one could imagine a role for this SPRY insertion in RNA binding (**see section 4.1.5**).

Previously, a preferential binding of DDX1 to polyA RNA sequences has been reported[109]. However, in the present study, no differences in binding affinities were observed between 10mer poly A, 10mer poly U or 13mer mixed sequence RNA in EMSAs (**Figure 3.3.3** and **suppl. Figure 7.1.10**). Furthermore, RNA molecules of different length and sequence, did stimulate nucleotide binding and hydrolysis to the same extent (**Figure 3.3.7**). This suggests, that the SPRY domain does not impose any sequence constraints on the substrate and uncomplexed DDX1 would therefore bind to almost any accessible oligonucleotide sequence.

The promiscuity in RNA sequence and length might even be essential for a DDX1 function as a general RNA processing enzyme. This has been noted for other DEAD-box proteins before - two prominent examples include Mss116 and eIF4AIII that both have to be promiscuous in RNA binding to fulfill their cellular function[19, 37]. The lack of specificity of DDX1 for certain RNA sequences could be easily overcome via recruitment to a specific target by interacting factors. Protein factors that e.g. interact with DDX1 in the HSPC117 complex could also constrain binding to specific structural elements in the substrate. Structural specificity has been described for DpbA related bacterial DEAD-box proteins, which recognize a RNA hairpin structure[272]. Consistent with a function of interacting factors in recruitment to specific substrates, the complex of DDX1 with CGI-99 and ASW has been found to be associated with cruciform DNA in pull-down experiments[113]. Noteworthy, in the gel-shift experiments, recombinant DDX1 alone did not bind to these DNAs and in addition, they did not influence ATPase activity (**suppl. Figure 7.1.16**).

These cruciform DNAs partially resemble the fold of the pre-tRNA (though they are in B-form and tRNA is in A-form[273]), the substrate of the DDX1-containing HSPC117 complex[114]. Thus, besides binding single stranded RNA without any sequence constraints, DDX1 might be targeted to specific structures by the HSPC117 complex.

Discussion

Irrespective of the binding to RNA substrates and stimulation of ATP hydrolysis by these interactions, the ATPase rate may be further influenced by additional factors as discussed in the following.

4.2.5 ATPase activity of DDX1 might be stimulated by external factors

The apparent rate of ATP hydrolysis by DDX1 is low, when compared to other DEAD-box helicases[56] (specifically in RNA bound form, **Table 4.1**). A low k_{cat} for DDX1 has been reported in another study before, where a discontinous assay for the quantification of ATP hydrolysis was used[135]. Concluding from the results of the coupled ATPase assay and the equilibrium titrations, the observed rate of hydrolysis may be limited by ADP release, which might be the rate limiting step of the reaction. The actual k_{cat} would then not be measured by simply detecting free ADP. A single turnover experiment could help to avoid the limitation by ADP release in the coupled ATPase assay and to determine the actual k_{cat} value. However, due to the low ATP affinity this is hardly possible.

Another reason for a low ATPase activity of DDX1 might be the formation of an auto-inhibited state. Such an autoinhibition was for instance described for DDX19, the human homologue of the Dbp5[28]. In the "open"-state, DDX19 is repressed by binding of an N-terminal α-helix between the two RecA-like domains[28] (**Figure 4.2.2**). This repressing element is part of an N-terminal extension of DDX19 that does not belong to the conserved helicase fold. If a similar autoinhibition mechanism applies to DDX1, the SPRY insertion is the prime candidate for an ATPase-repressing domain.

The DEAD-box protein DDX19/Dbp5, but also eIF4A, is specifically stimulated in its ATP hydrolysis activity by binding of the co-factors Gle1-IP$_6$ and eIF4G, respectively[19]. These factors activate their DEAD-box partner by stimulating RNA release[62] (**Figure 4.2.2**). For Dbp5, the ATPase stimulating function of Gle1-IP$_6$ is further coupled to ADP-exchange by Nup159[67]. Similarly, in DDX1 external factors may also regulate both, RNA release and ADP exchange.

In general, ATP hydrolysis is not the main enzymatic activity of DEAD-box proteins, but rather means to support substrate unwinding[18]. Whereas hydrolysis of ATP could be measured in the experiments performed during this thesis, unwinding of double-stranded oligonucleotides by DDX1 was not observed. This enzymatic deficiency will be discussed in the next section.

Discussion

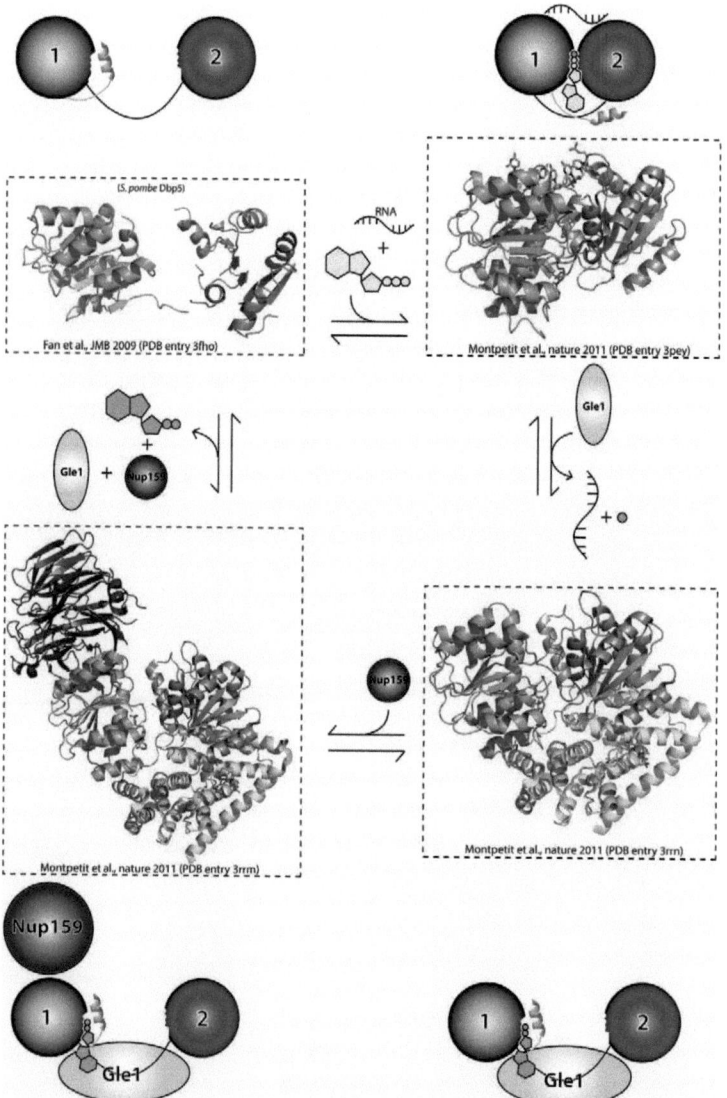

Figure 4.2.2 **The ATPase cycle of Dbp5 is regulated by an RNA release- and a nucleotide exchange-factor.** The DEAD-box protein Dbp5 (DDX19 in *H. sapiens*) displays a slow intrinsic ATPase activity due to an α-helical extension that binds to the ATPase site in the "open"-state of the helicase. The ATPase cycle is accelerated by binding of the co-factors Gle1- InsP$_6$ and Nup159. Gle1-InsP$_6$ activates the ATPase reaction by stimulating RNA release. The nucleoporin Nup159 (Nup214 in *H. sapiens*) acts as a

nucleotide exchange factor to remove ADP after hydrolysis. It is conceivable that DDX1 requires similar factors to stimulate its enzymatic activity, since for DDX1 ADP affinity is exceptionally high and the basal ATPase rate is slow. The factors of the pentameric HSPC117 complex are candidates for such activities. Structures of different enzymatic states are depicted that have been accessible by X-ray crystallography. Apo Dbp5 is in the "open"-state of the helicase(upper left) and forms the "closed"-state upon binding of RNA and ATP (upper right). Binding of Gle1 after ATP-hydrolysis leads to a separation of both RecA-like domains and RNA release (lower right). Finally Nup159 binds and acts as an ADP-exchange-factor, which helps to reset the enzyme (lower left). All structures except for the apo-form represent *S. cerevisiae* proteins. Inositol hexakisphosphate (InsP$_6$) acts as a small molecular tether for the Gle1-Dbp5 interaction (represented as stick model in the structures and omitted from the cartoon representation).
Note that in the nucleotide- and co-factor-free state, RecA-like domains 1 and 2 display high flexibility in their orientation. In the crystal structure of the apo-form "open"-state, the distance between both domains is larger than in the defined "open"-state with Gle1 bound.

4.2.6 Helicase deficiency of DDX1 hints towards a clamping function

Recombinant DDX1 clearly showed an ATP hydrolysis activity (**Figure 3.3.10**), however, in the gel-based helicase assay recombinant DDX1 was unable to separate DNA/RNA or RNA/RNA substrates (**Figure 3.3.11**). This is in contrast to previous reports from the lab of Roseline Godbout that has shown efficient unwinding of relatively long RNA or DNA/RNA hybrid substrates[111]. The same oligonucleotide substrates as described in these studies were tested, but could not be separated by DDX1 during this study. These substrates harbor long stretches of 29 paired nucleotides, which is too long to be efficiently processed by most DEAD-box helicases that commonly do not unwind duplexes, exceeding one-and-a-half helical turns[157]. To exclude that the substrate is not unwound by DDX1 due to processivity problems, shorter double-stranded oligonucleotides were tested. These shorter oligonucleotides have a 13-bp duplex region and have been used as model substrates for several DEAD-box proteins[49]. Nevertheless, no unwinding could be observed for the oligonucleotides with shorter duplex region (**Figure 3.3.12**). Interestingly, another study suggested that isolated recombinant DDX1 does not exhibit any RNA unwinding activity and only exerts its function in protein complexes that were co-immunoprecipitated with DDX1[109].

Although, it can not be completely excluded that the FAM- or the Fl-label hinders unwinding, positioning at the 3'- or 5'-end of either DNA or RNA substrate did not make any difference in the displacement reaction. In the gel-shift experiments DDX1 did bind to ssRNA despite the presence of a FAM-label. Additionally the label is comparably small (380 Da) and DEAD-box helicases contact only ~ 10 nucleotides[274]. Since all double-stranded oligonucleotides were much longer, DDX1 could simply bind to a part of the potential substrate that does not impose any sterical hindrance.

Since unspecific RNA binding was observed in EMSA- and equilibrium titration experiments (**Figure 3.3.3** and **3.3.7**), it is unlikely that the sequence of the double-stranded substrate is one of the factors that prevented unwinding. Furthermore, all assays for measuring the unwinding activity of DEAD-box proteins that have been described in the literature so far, assume sequence independent unwinding[49, 157, 275]. The construct DDX1-728 used in the assays, lacks only 12 residues from the C-terminus when compared to full-length DDX1 (**Figure 3.1.1**). Most probably these residues that belong to a C-terminal

Discussion

extension (that is not part of the helicase core) are not essential for the enzymatic activity as they are completely unconserved (in contrast to the rest of the protein, see **Figure 3.1.1**). Moreover for many DEAD-box proteins, it was shown that constructs, truncated down to the basic RecA-like helicase core are active in dsRNA unwinding[31, 54, 276].

The failure in detecting for separation of double-stranded substrate by the DDX1 construct (as used in this thesis) (**Figure 3.3.11**) could be due to missing co-factors or incorrect folding of the recombinant protein. Binding of adenosine-nucleotides and RNA could be measured and affinity values were in the range of what has been reported for other DEAD-box proteins[57, 58] (**Table 4.1**). This suggests that residues, essential for ATP binding, especially the P-loop, must be in the correct conformation. Furthermore, the RNA binding surface on top of RecA-like domains 1 and 2 is almost certainly correctly assembled. Thermal stability of DDX1 was in the range of stable proteins[277] and ATP hydrolysis activity could be observed (**Figure 3.3.1 and 3.3.10**). All those results indicate that the recombinant protein is most likely properly folded. This leaves only the lack of co-factors as a possible cause of the helicase deficiency, which has been suggested in another report before[109].

On the other hand, it is possible that the basic function of DDX1 is RNA clamping and that it serves as a structural scaffold for the assembly of larger protein complexes. For instance, DDX1 could help to position the HSPC117 complex components on tRNA substrate. Such an assembly function has e.g. been described for the DEAD-box proteins Dbp5 and eIF4AIII that both clamp on mRNA substrates[23-25]. In DDX1, the observed pronounced preference for ADP over ATP supports a scaffolding function (**Figure 3.3.5**). As mentioned before, DDX1 may need an activating nucleotide exchange factor for ADP release. In the absence of such a factor, as it was the case in the helicase assay, DDX1 remains bound to ADP and RNA. Local activation by co-factors has been reported for the "clampers" Dbp5[25, 278] and eIF4AIII[279] (**Figure 4.2.2**).

The DEAD-box protein DED1 (**suppl. Figure 7.3.12**) that has been used as a positive control in the displacement assay, nicely illustrates the functional differences between a processive helicase and a clamping, scaffolding enzyme. During translation initiation the model DEAD-box protein eIF4A clamps onto the 5'-end of the mRNA and helps to assemble the components of the cap-binding complex[280], until it is locally activated by eIF4G[61, 62]. eIF4A only serves a scaffolding function and it is not capable of unwinding secondary structures in the 5' untranslated region (UTR) of the mRNA. The more processive helicase DED1 is required to unwind these secondary structures in the 5' UTR to facilitate ribosome scanning[281]. Likewise DDX1 may assemble the components of the tRNA processing complex[114] on the substrate, but leave the actual "work" to the enzymatically active HSPC117.

Discussion

4.3 Conclusion and Outlook

4.3.1 DDX1 is a novel, human DEAD-box protein involved in a plethora of cellular functions

In this thesis the human DEAD box protein 1, DDX1 was characterized both structurally and functionally. DDX1 differs from other proteins of its helicase-family by a large insertion in the conserved helicase core. This work verified that the insertion corresponds to a SPRY domain. In other proteins, SPRY domains interact with the DEAD-box helicase core of a different protein[100, 225, 226] and DDX1 seems to be the first example of a DEAD-box protein with a "built-in" SPRY domain. DDX1 was found to be conserved from *H. sapiens* to *C. elegans*, but no yeast-homolog could be detected. Previous studies sparked high interest in DDX1, since it was shown to be involved in mRNA- and tRNA- processing and to play an important role in human pathogenesis as it is amplified in tumor tissue[91] and is an essential host factor for HIV-1 replication[135].

Here, a variant of DDX1 was designed that can be expressed in bacterial cells and offers an easy and fast access to recombinant material. Experiments showed that DDX1 is not only structurally distinct from other DEAD-box proteins, but also its enzymatic parameters are very different when compared to reported values of other DEAD-box proteins[60]. The established protocols for the preparation of recombinant DDX1 can be used for future studies on the HSPC117 complex and its role in tRNA processing.

4.3.2 Structural characterization of the SPRY domain

The SPRY domain was crystallized and its structure was determined to near atomic resolution. This represents the first structural information of DDX1 and the HSPC117 tRNA ligation complex in general. The SPRY structure shows two layers of concave shaped β-sheets that stack together to form a compact β-sandwich conformation and a third small β-sheet that forms a lid over the β-sandwich core. In comparison with complex structures of other mammalian SPRY domains, it was observed that the conformation of previously described interaction loops[101, 105] is different in the DDX1 SPRY domain and residues located at the canonical interaction surface A are not conserved. Conserved, however, was a positively charged surface patch that differs from the canonical interaction surface. The high conservation combined with the charge distribution, suggests that this surface patch may mediate peptide-binding or could potentially also enlarge the RNA-interaction surface of DDX1. The SPRY domain is the putative element that establishes the interaction of DDX1 with HIV-1 Rev protein as it was shown that the Rev-binding domain is contained entirely within the N-terminal region of DDX1[135]. Therefore, the SPRY structure provides the basis for further studies on the role of DDX1 in HIV-1 replication and the detailed biochemical characterization of the SPRY interaction platform and the recruitment to the Rev protein.

4.3.3 Functional characterization of DDX1

DDX1 displayed an ATP affinity, similar to other DEAD-box proteins, but showed an unusually tight ADP binding. The difference in the affinities between adenosine-di- and triphosphates that was determined is so profound that in the cytoplasm it would lead to "ADP-stalling" and inactivate DDX1. This suggests that an nucleotide exchange factor as described previously for DDX19[67], is potentially required for ADP release. External co-factors of the HSPC117 complex may take over this function and by acting as nucleotide exchanger could also direct spatial activation and distribution of DDX1. Despite an experimentally verified ability to hydrolyze ATP, recombinant DDX1 was helicase deficient, in contrast to the yeast DEAD-box helicase DED1. This deficiency hints again to the requirement of co-factors and a function of DDX1 in substrate clamping. DDX1 did not display RNA-sequence specificity.

A tight coupling between the binding of RNA and ATP to DDX1 was observed across all experiments. Cooperativity in RNA and ATP binding can be explained by a model of DDX1 in which both substrates can induce the conformational transition to the "closed"-state of the helicase by their own. This model is corroborated by the observed intrinsic ATPase activity, which requires a "closed"-state.

The kinetic data extend the working model for eukaryotic DEAD-box proteins by introducing synergistic effects of ATP and RNA binding that have not been observed to such an extent before. Further *in vivo* experiments can make use of the cooperativity model and the reported affinity and binding constants for ATP, ADP and RNA for an in-depth analysis for the hijacking function of viral transcripts and their modulatory propensity.

5. Acknowledgements

This work would have hardly been possible without the continous support, aid, help, backup and encouragement of numerous people during the last years. I am thankful to the core to everyone who contributed in the preparation, planing and completion of this thesis.

I am deep in the debt to so many people and I can only try to name some of the most influential of them here, though this list will be far from complete.

I would like to thank...

... my direct supervisor Dr. Anton Meinhart for accepting me as a PhD student and giving me the opportunity to work on this exciting and challenging project, for his scientific advices, many helpful discussions, his guidance in the tough times and his help for finding my way to the right track. Also for critically and thoroughly reading this thesis.

... PD Dr. Jochen Reinstein for all the help, scientific advice, discussions and support in data fitting, for having an open door for questions, his valuable suggestions and all the counseling when it comes to kinetics.

... Prof. Dr. Ilme Schlichting for her continuous support, for invaluable advice, both scientific and personal and for accompanying me through the TAC meetings.

... Prof. Dr. Oliver Gruss and Dr. Aurelio Teleman for a critical evaluation of this thesis and for assessing my disputation.

... Dr. Thomas Barends for help with crystallographic questions, entertaining movie citations and surgical crystal fishing.

... all present and former members of the Meinhart lab for scientific and practical support, for the enjoyable and friendly atmosphere in the lab and the pleasant climate throughout the years. Thanks to Andrea, Brad, Christina, Christian, Florence, Hannes, Iris, Juliane and Stefano. I would like to thank Maike Gebhardt for her kindness and great assistance. I would like to thank Aytac Dikfidan for continuous support and many enjoyable scientific and non-scientific discussions. Also for all his help with pymol-scripting and Linux-operation.

... Chris Roome for excellent IT service and his discreet kindness.

Acknowledgements

…the SLS Team Dortmund-Heidelberg for collection, transfer and processing of data.

… present and former members of the BMM department for the great working atmosphere. Stephan for supplying fruits for healthy diet on the afternoon and for help with plotting of the global fit data.

… people in my office for being open for questions and coming up with advice on scientifc, computer and private issues. Thanks to Kristina for the enjoyable atmosphere and to Mirek for helping with crystallographic problems.

… Melanie Müller and Marion Gradl for excellent support in all MS-things.

… the Mensa-gang, Cathleen, Udo, Rob, Diana and many more for enjoyable luch-times.[197]

… all the practical project students for keeping the lab running.

… Nicolas Werbeck for supervising my master-thesis and helping me to acquire the basic knowledge of all kinetic-related things.

… many people outside the MPI, that helped me getting through the rainy days of research. Special thanks to Conny, Wolf, Aniko and Tabea.

I am very grateful to my parents that always supported me, no matter what and have to suffer from my continuous negligence.

6. References

[1] A. K. Byrd, K. D. Raney, *Superfamily 2 helicases*, Front Biosci **2012**, *17*, 2070.
[2] D. Lane, *Enlarged family of putative helicases*, Nature **1988**, *334*, 478.
[3] T. C. Hodgman, *A new superfamily of replicative proteins*, Nature **1988**, *333*, 22.
[4] K. Geider, H. Hoffmann-Berling, *Proteins controlling the helical structure of DNA*, Annu Rev Biochem **1981**, *50*, 233.
[5] H. Hoffmann-Berling, *DNA unwinding enzymes*, Prog Clin Biol Res **1982**, *102 Pt C*, 89.
[6] K. E. Gorbalenya AE, *Helicases: amino acid comparisons and structure-function relationships*, Curr Opin Struct Biol **1993**, *3*, 419.
[7] M. R. Singleton, M. S. Dillingham, D. B. Wigley, *Structure and mechanism of helicases and nucleic acid translocases*, Annu Rev Biochem **2007**, *76*, 23.
[8] J. E. Walker, M. Saraste, M. J. Runswick, N. J. Gay, *Distantly related sequences in the alpha- and beta-subunits of ATP synthase, myosin, kinases and other ATP-requiring enzymes and a common nucleotide binding fold*, EMBO J **1982**, *1*, 945.
[9] M. Saraste, P. R. Sibbald, A. Wittinghofer, *The P-loop--a common motif in ATP- and GTP-binding proteins*, Trends Biochem Sci **1990**, *15*, 430.
[10] A. E. Gorbalenya, E. V. Koonin, A. P. Donchenko, V. M. Blinov, *Two related superfamilies of putative helicases involved in replication, recombination, repair and expression of DNA and RNA genomes*, Nucleic Acids Res **1989**, *17*, 4713.
[11] A. M. Pyle, *Translocation and unwinding mechanisms of RNA and DNA helicases*, Annu Rev Biophys **2008**, *37*, 317.
[12] M. E. Fairman-Williams, U. P. Guenther, E. Jankowsky, *SF1 and SF2 helicases: family matters*, Curr Opin Struct Biol **2010**, *20*, 313.
[13] E. Jankowsky, M. E. Fairman, *RNA helicases--one fold for many functions*, Curr Opin Struct Biol **2007**, *17*, 316.
[14] N. K. Tanner, O. Cordin, J. Banroques, M. Doere, P. Linder, *The Q motif: a newly identified motif in DEAD box helicases may regulate ATP binding and hydrolysis*, Mol Cell **2003**, *11*, 127.
[15] J. M. Caruthers, D. B. McKay, *Helicase structure and mechanism*, Curr Opin Struct Biol **2002**, *12*, 123.
[16] H. S. Subramanya, L. E. Bird, J. A. Brannigan, D. B. Wigley, *Crystal structure of a DExx box DNA helicase*, Nature **1996**, *384*, 379.
[17] P. Linder, P. F. Lasko, M. Ashburner, P. Leroy, P. J. Nielsen, K. Nishi, J. Schnier, P. P. Slonimski, *Birth of the D-E-A-D box*, Nature **1989**, *337*, 121.
[18] O. Cordin, J. Banroques, N. K. Tanner, P. Linder, *The DEAD-box protein family of RNA helicases*, Gene **2006**, *367*, 17.
[19] P. Linder, E. Jankowsky, *From unwinding to clamping - the DEAD box RNA helicase family*, Nat Rev Mol Cell Biol **2011**, *12*, 505.
[20] C. Pan, R. Russell, *Roles of DEAD-box proteins in RNA and RNP Folding*, RNA Biol **2010**, *7*, 667.
[21] B. K. Ray, T. G. Lawson, J. C. Kramer, M. H. Cladaras, J. A. Grifo, R. D. Abramson, W. C. Merrick, R. E. Thach, *ATP-dependent unwinding of messenger RNA structure by eukaryotic initiation factors*, J Biol Chem **1985**, *260*, 7651.
[22] J. de la Cruz, D. Kressler, P. Linder, *Unwinding RNA in Saccharomyces cerevisiae: DEAD-box proteins and related families*, Trends Biochem Sci **1999**, *24*, 192.
[23] C. B. Andersen, L. Ballut, J. S. Johansen, H. Chamieh, K. H. Nielsen, C. L. Oliveira, J. S. Pedersen, B. Seraphin, H. Le Hir, G. R. Andersen, *Structure of the exon junction core complex with a trapped DEAD-box ATPase bound to RNA*, Science **2006**, *313*, 1968.
[24] F. Bono, J. Ebert, E. Lorentzen, E. Conti, *The crystal structure of the exon junction complex reveals how it maintains a stable grip on mRNA*, Cell **2006**, *126*, 713.

[25] B. Montpetit, N. D. Thomsen, K. J. Helmke, M. A. Seeliger, J. M. Berger, K. Weis, *A conserved mechanism of DEAD-box ATPase activation by nucleoporins and InsP6 in mRNA export*, Nature **2011**, *472*, 238.

[26] H. Shi, O. Cordin, C. M. Minder, P. Linder, R. M. Xu, *Crystal structure of the human ATP-dependent splicing and export factor UAP56*, Proc Natl Acad Sci U S A **2004**, *101*, 17628.

[27] Z. Cheng, J. Coller, R. Parker, H. Song, *Crystal structure and functional analysis of DEAD-box protein Dhh1p*, RNA **2005**, *11*, 1258.

[28] R. Collins, T. Karlberg, L. Lehtio, P. Schutz, S. van den Berg, L. G. Dahlgren, M. Hammarstrom, J. Weigelt, H. Schuler, *The DEXD/H-box RNA helicase DDX19 is regulated by an {alpha}-helical switch*, J Biol Chem **2009**, *284*, 10296.

[29] J. Benz, H. Trachsel, U. Baumann, *Crystal structure of the ATPase domain of translation initiation factor 4A from Saccharomyces cerevisiae--the prototype of the DEAD box protein family*, Structure **1999**, *7*, 671.

[30] A. Pause, N. Sonenberg, *Mutational analysis of a DEAD box RNA helicase: the mammalian translation initiation factor eIF-4A*, EMBO J **1992**, *11*, 2643.

[31] T. Sengoku, O. Nureki, A. Nakamura, S. Kobayashi, S. Yokoyama, *Structural basis for RNA unwinding by the DEAD-box protein Drosophila Vasa*, Cell **2006**, *125*, 287.

[32] P. Linder, P. Lasko, *Bent out of shape: RNA unwinding by the DEAD-box helicase Vasa*, Cell **2006**, *125*, 219.

[33] R. M. Story, H. Li, J. N. Abelson, *Crystal structure of a DEAD box protein from the hyperthermophile Methanococcus jannaschii*, Proc Natl Acad Sci U S A **2001**, *98*, 1465.

[34] M. Hilbert, A. R. Karow, D. Klostermeier, *The mechanism of ATP-dependent RNA unwinding by DEAD box proteins*, Biol Chem **2009**, *390*, 1237.

[35] M. Del Campo, A. M. Lambowitz, *Structure of the Yeast DEAD box protein Mss116p reveals two wedges that crimp RNA*, Mol Cell **2009**, *35*, 598.

[36] A. L. Mallam, M. Del Campo, B. Gilman, D. J. Sidote, A. M. Lambowitz, *Structural basis for RNA-duplex recognition and unwinding by the DEAD-box helicase Mss116p*, Nature **2012**, *490*, 121.

[37] A. A. Putnam, E. Jankowsky, *DEAD-box helicases as integrators of RNA, nucleotide and protein binding*, Biochim Biophys Acta **2013**, *1829*, 884.

[38] Z. Cheng, D. Muhlrad, M. K. Lim, R. Parker, H. Song, *Structural and functional insights into the human Upf1 helicase core*, EMBO J **2007**, *26*, 253.

[39] J. W. Hardin, Y. X. Hu, D. B. McKay, *Structure of the RNA binding domain of a DEAD-box helicase bound to its ribosomal RNA target reveals a novel mode of recognition by an RNA recognition motif*, J Mol Biol **2010**, *402*, 412.

[40] D. Klostermeier, M. G. Rudolph, *A novel dimerization motif in the C-terminal domain of the Thermus thermophilus DEAD box helicase Hera confers substantial flexibility*, Nucleic Acids Res **2009**, *37*, 421.

[41] M. G. Rudolph, D. Klostermeier, *The Thermus thermophilus DEAD box helicase Hera contains a modified RNA recognition motif domain loosely connected to the helicase core*, RNA **2009**, *15*, 1993.

[42] L. Steimer, J. P. Wurm, M. H. Linden, M. G. Rudolph, J. Wohnert, D. Klostermeier, *Recognition of two distinct elements in the RNA substrate by the RNA-binding domain of the T. thermophilus DEAD box helicase Hera*, Nucleic Acids Res **2013**, *41*, 6259.

[43] W. Breitwieser, F. H. Markussen, H. Horstmann, A. Ephrussi, *Oskar protein interaction with Vasa represents an essential step in polar granule assembly*, Genes Dev **1996**, *10*, 2179.

[44] S. Styhler, A. Nakamura, P. Lasko, *VASA localization requires the SPRY-domain and SOCS-box containing protein, GUSTAVUS*, Dev Cell **2002**, *3*, 865.

[45] M. G. Rudolph, R. Heissmann, J. G. Wittmann, D. Klostermeier, *Crystal structure and nucleotide binding of the Thermus thermophilus RNA helicase Hera N-terminal domain*, J Mol Biol **2006**, *361*, 731.

[46] P. Emsley, B. Lohkamp, W. G. Scott, K. Cowtan, *Features and development of Coot*, Acta Crystallogr D Biol Crystallogr **2010**, *66*, 486.

References

[47] D. Klostermeier, *Rearranging RNA structures at 75 degrees C? toward the molecular mechanism and physiological function of the thermus thermophilus DEAD-box helicase hera*, Biopolymers **2013**, *99*, 1137.
[48] J. Napetschnig, S. A. Kassube, E. W. Debler, R. W. Wong, G. Blobel, A. Hoelz, *Structural and functional analysis of the interaction between the nucleoporin Nup214 and the DEAD-box helicase Ddx19*, Proc Natl Acad Sci U S A **2009**, *106*, 3089.
[49] E. Jankowsky, A. Putnam, *Duplex unwinding with DEAD-box proteins*, Methods Mol Biol **2010**, *587*, 245.
[50] T. Bizebard, I. Ferlenghi, I. Iost, M. Dreyfus, *Studies on three E. coli DEAD-box helicases point to an unwinding mechanism different from that of model DNA helicases*, Biochemistry **2004**, *43*, 7857.
[51] Q. Yang, E. Jankowsky, *The DEAD-box protein Ded1 unwinds RNA duplexes by a mode distinct from translocating helicases*, Nat Struct Mol Biol **2006**, *13*, 981.
[52] Q. Yang, M. Del Campo, A. M. Lambowitz, E. Jankowsky, *DEAD-box proteins unwind duplexes by local strand separation*, Mol Cell **2007**, *28*, 253.
[53] G. W. Rogers, Jr., N. J. Richter, W. C. Merrick, *Biochemical and kinetic characterization of the RNA helicase activity of eukaryotic initiation factor 4A*, J Biol Chem **1999**, *274*, 12236.
[54] G. W. Rogers, Jr., W. F. Lima, W. C. Merrick, *Further characterization of the helicase activity of eIF4A. Substrate specificity*, J Biol Chem **2001**, *276*, 12598.
[55] G. W. Rogers, Jr., N. J. Richter, W. F. Lima, W. C. Merrick, *Modulation of the helicase activity of eIF4A by eIF4B, eIF4H, and eIF4F*, J Biol Chem **2001**, *276*, 30914.
[56] I. Garcia, O. C. Uhlenbeck, *Differential RNA-dependent ATPase activities of four rRNA processing yeast DEAD-box proteins*, Biochemistry **2008**, *47*, 12562.
[57] A. Henn, W. Cao, D. D. Hackney, E. M. De La Cruz, *The ATPase cycle mechanism of the DEAD-box rRNA helicase, DbpA*, J Mol Biol **2008**, *377*, 193.
[58] W. Cao, M. M. Coman, S. Ding, A. Henn, E. R. Middleton, M. J. Bradley, E. Rhoades, D. D. Hackney, A. M. Pyle, E. M. De La Cruz, *Mechanism of Mss116 ATPase reveals functional diversity of DEAD-Box proteins*, J Mol Biol **2011**, *409*, 399.
[59] I. Iost, M. Dreyfus, P. Linder, *Ded1p, a DEAD-box protein required for translation initiation in Saccharomyces cerevisiae, is an RNA helicase*, J Biol Chem **1999**, *274*, 17677.
[60] A. Henn, M. J. Bradley, E. M. De La Cruz, *ATP utilization and RNA conformational rearrangement by DEAD-box proteins*, Annu Rev Biophys **2012**, *41*, 247.
[61] P. Schutz, M. Bumann, A. E. Oberholzer, C. Bieniossek, H. Trachsel, M. Altmann, U. Baumann, *Crystal structure of the yeast eIF4A-eIF4G complex: an RNA-helicase controlled by protein-protein interactions*, Proc Natl Acad Sci U S A **2008**, *105*, 9564.
[62] M. Hilbert, F. Kebbel, A. Gubaev, D. Klostermeier, *eIF4G stimulates the activity of the DEAD box protein eIF4A by a conformational guidance mechanism*, Nucleic Acids Res **2011**, *39*, 2260.
[63] P. G. Loh, H. S. Yang, M. A. Walsh, Q. Wang, X. Wang, Z. Cheng, D. Liu, H. Song, *Structural basis for translational inhibition by the tumour suppressor Pdcd4*, EMBO J **2009**, *28*, 274.
[64] J. H. Chang, Y. H. Cho, S. Y. Sohn, J. M. Choi, A. Kim, Y. C. Kim, S. K. Jang, Y. Cho, *Crystal structure of the eIF4A-PDCD4 complex*, Proc Natl Acad Sci U S A **2009**, *106*, 3148.
[65] J. M. Caruthers, E. R. Johnson, D. B. McKay, *Crystal structure of yeast initiation factor 4A, a DEAD-box RNA helicase*, Proc Natl Acad Sci U S A **2000**, *97*, 13080.
[66] H. S. Yang, A. P. Jansen, A. A. Komar, X. Zheng, W. C. Merrick, S. Costes, S. J. Lockett, N. Sonenberg, N. H. Colburn, *The transformation suppressor Pdcd4 is a novel eukaryotic translation initiation factor 4A binding protein that inhibits translation*, Mol Cell Biol **2003**, *23*, 26.
[67] K. N. Noble, E. J. Tran, A. R. Alcazar-Roman, C. A. Hodge, C. N. Cole, S. R. Wente, *The Dbp5 cycle at the nuclear pore complex during mRNA export II: nucleotide cycling and mRNP remodeling by Dbp5 are controlled by Nup159 and Gle1*, Genes Dev **2011**, *25*, 1065.
[68] K. H. Nielsen, H. Chamieh, C. B. Andersen, F. Fredslund, K. Hamborg, H. Le Hir, G. R. Andersen, *Mechanism of ATP turnover inhibition in the EJC*, RNA **2009**, *15*, 67.
[69] P. Soultanas, D. B. Wigley, *DNA helicases: 'inching forward'*, Curr Opin Struct Biol **2000**, *10*, 124.

[70] P. Soultanas, D. B. Wigley, Unwinding the 'Gordian knot' of helicase action, Trends Biochem Sci **2001**, 26, 47.
[71] Y. Chen, J. P. Potratz, P. Tijerina, M. Del Campo, A. M. Lambowitz, R. Russell, DEAD-box proteins can completely separate an RNA duplex using a single ATP, Proc Natl Acad Sci U S A **2008**, 105, 20203.
[72] A. Henn, W. Cao, N. Licciardello, S. E. Heitkamp, D. D. Hackney, E. M. De La Cruz, Pathway of ATP utilization and duplex rRNA unwinding by the DEAD-box helicase, DbpA, Proc Natl Acad Sci U S A **2010**, 107, 4046.
[73] F. Liu, A. Putnam, E. Jankowsky, ATP hydrolysis is required for DEAD-box protein recycling but not for duplex unwinding, Proc Natl Acad Sci U S A **2008**, 105, 20209.
[74] R. Aregger, D. Klostermeier, The DEAD box helicase YxiN maintains a closed conformation during ATP hydrolysis, Biochemistry **2009**, 48, 10679.
[75] J. R. Lorsch, D. Herschlag, The DEAD box protein eIF4A. 1. A minimal kinetic and thermodynamic framework reveals coupled binding of RNA and nucleotide, Biochemistry **1998**, 37, 2180.
[76] M. H. Linden, R. K. Hartmann, D. Klostermeier, The putative RNase P motif in the DEAD box helicase Hera is dispensable for efficient interaction with RNA and helicase activity, Nucleic Acids Res **2008**, 36, 5800.
[77] A. R. Karow, D. Klostermeier, A conformational change in the helicase core is necessary but not sufficient for RNA unwinding by the DEAD box helicase YxiN, Nucleic Acids Res **2009**, 37, 4464.
[78] K. J. Polach, O. C. Uhlenbeck, Cooperative binding of ATP and RNA substrates to the DEAD/H protein DbpA, Biochemistry **2002**, 41, 3693.
[79] J. K. Grohman, M. Del Campo, H. Bhaskaran, P. Tijerina, A. M. Lambowitz, R. Russell, Probing the mechanisms of DEAD-box proteins as general RNA chaperones: the C-terminal domain of CYT-19 mediates general recognition of RNA, Biochemistry **2007**, 46, 3013.
[80] B. Theissen, A. R. Karow, J. Kohler, A. Gubaev, D. Klostermeier, Cooperative binding of ATP and RNA induces a closed conformation in a DEAD box RNA helicase, Proc Natl Acad Sci U S A **2008**, 105, 548.
[81] M. E. Fairman, P. A. Maroney, W. Wang, H. A. Bowers, P. Gollnick, T. W. Nilsen, E. Jankowsky, Protein displacement by DExH/D "RNA helicases" without duplex unwinding, Science **2004**, 304, 730.
[82] E. Jankowsky, H. Bowers, Remodeling of ribonucleoprotein complexes with DExH/D RNA helicases, Nucleic Acids Res **2006**, 34, 4181.
[83] M. K. Lund, C. Guthrie, The DEAD-box protein Dbp5p is required to dissociate Mex67p from exported mRNPs at the nuclear rim, Mol Cell **2005**, 20, 645.
[84] Q. Yang, E. Jankowsky, ATP- and ADP-dependent modulation of RNA unwinding and strand annealing activities by the DEAD-box protein DED1, Biochemistry **2005**, 44, 13591.
[85] C. Halls, S. Mohr, M. Del Campo, Q. Yang, E. Jankowsky, A. M. Lambowitz, Involvement of DEAD-box proteins in group I and group II intron splicing. Biochemical characterization of Mss116p, ATP hydrolysis-dependent and -independent mechanisms, and general RNA chaperone activity, J Mol Biol **2007**, 365, 835.
[86] Q. Yang, M. E. Fairman, E. Jankowsky, DEAD-box-protein-assisted RNA structure conversion towards and against thermodynamic equilibrium values, J Mol Biol **2007**, 368, 1087.
[87] B. C. Valdez, Structural domains involved in the RNA folding activity of RNA helicase II/Gu protein, Eur J Biochem **2000**, 267, 6395.
[88] H. Bhaskaran, R. Russell, Kinetic redistribution of native and misfolded RNAs by a DEAD-box chaperone, Nature **2007**, 449, 1014.
[89] S. Mohr, J. M. Stryker, A. M. Lambowitz, A DEAD-box protein functions as an ATP-dependent RNA chaperone in group I intron splicing, Cell **2002**, 109, 769.
[90] F. Liu, A. A. Putnam, E. Jankowsky, DEAD-Box Helicases Form Nucleotide-Dependent, Long-Lived Complexes with RNA, Biochemistry **2014**.
[91] R. Godbout, M. Packer, W. Bie, Overexpression of a DEAD box protein (DDX1) in neuroblastoma and retinoblastoma cell lines, J Biol Chem **1998**, 273, 21161.

References

[92] R. Godbout, M. Hale, D. Bisgrove, *A human DEAD box protein with partial homology to heterogeneous nuclear ribonucleoprotein U*, Gene **1994**, *138*, 243.
[93] J. Fang, S. Kubota, B. Yang, N. Zhou, H. Zhang, R. Godbout, R. J. Pomerantz, *A DEAD box protein facilitates HIV-1 replication as a cellular co-factor of Rev*, Virology **2004**, *330*, 471.
[94] R. Godbout, L. Li, R. Z. Liu, K. Roy, *Role of DEAD box 1 in retinoblastoma and neuroblastoma*, Future Oncol **2007**, *3*, 575.
[95] K. E. Zinsmaier, K. K. Eberle, E. Buchner, N. Walter, S. Benzer, *Paralysis and early death in cysteine string protein mutants of Drosophila*, Science **1994**, *263*, 977.
[96] S. R. Schmid, P. Linder, *D-E-A-D protein family of putative RNA helicases*, Mol Microbiol **1992**, *6*, 283.
[97] C. Ponting, J. Schultz, P. Bork, *SPRY domains in ryanodine receptors (Ca(2+)-release channels)*, Trends Biochem Sci **1997**, *22*, 193.
[98] A. A. D'Cruz, J. J. Babon, R. S. Norton, N. A. Nicola, S. E. Nicholson, *Structure and function of the SPRY/B30.2 domain proteins involved in innate immunity*, Protein Sci **2012**, *22*, 1.
[99] D. A. Rhodes, B. de Bono, J. Trowsdale, *Relationship between SPRY and B30.2 protein domains. Evolution of a component of immune defence?*, Immunology **2005**, *116*, 411.
[100] J. S. Woo, J. H. Imm, C. K. Min, K. J. Kim, S. S. Cha, B. H. Oh, *Structural and functional insights into the B30.2/SPRY domain*, EMBO J **2006**, *25*, 1353.
[101] J. S. Woo, H. Y. Suh, S. Y. Park, B. H. Oh, *Structural basis for protein recognition by B30.2/SPRY domains*, Mol Cell **2006**, *24*, 967.
[102] D. A. Rhodes, G. Ihrke, A. T. Reinicke, G. Malcherek, M. Towey, D. A. Isenberg, J. Trowsdale, *The 52 000 MW Ro/SS-A autoantigen in Sjogren's syndrome/systemic lupus erythematosus (Ro52) is an interferon-gamma inducible tripartite motif protein associated with membrane proximal structures*, Immunology **2002**, *106*, 246.
[103] D. Wang, Z. Li, E. M. Messing, G. Wu, *Activation of Ras/Erk pathway by a novel MET-interacting protein RanBPM*, J Biol Chem **2002**, *277*, 36216.
[104] T. Niikura, Y. Hashimoto, H. Tajima, M. Ishizaka, Y. Yamagishi, M. Kawasumi, M. Nawa, K. Terashita, S. Aiso, I. Nishimoto, *A tripartite motif protein TRIM11 binds and destabilizes Humanin, a neuroprotective peptide against Alzheimer's disease-relevant insults*, Eur J Neurosci **2003**, *17*, 1150.
[105] P. Filippakopoulos, A. Low, T. D. Sharpe, J. Uppenberg, S. Yao, Z. Kuang, P. Savitsky, R. S. Lewis, S. E. Nicholson, R. S. Norton, A. N. Bullock, *Structural basis for Par-4 recognition by the SPRY domain- and SOCS box-containing proteins SPSB1, SPSB2, and SPSB4*, J Mol Biol **2010**, *401*, 389.
[106] Z. Kuang, S. Yao, Y. Xu, R. S. Lewis, A. Low, S. L. Masters, T. A. Willson, T. B. Kolesnik, S. E. Nicholson, T. J. Garrett, R. S. Norton, *SPRY domain-containing SOCS box protein 2: crystal structure and residues critical for protein binding*, J Mol Biol **2009**, *386*, 662.
[107] Y. Chen, F. Cao, B. Wan, Y. Dou, M. Lei, *Structure of the SPRY domain of human Ash2L and its interactions with RbBP5 and DPY30*, Cell Res **2012**, *22*, 598.
[108] S. Trowitzsch, *Functional and structural investigation of spliceosomal snRNPs*, Dissertation - Uni Göttingen **2008**, university of Goettingen, 194.
[109] H. C. Chen, W. C. Lin, Y. G. Tsay, S. Lee, C. J. Chang, *An RNA helicase, DDX1, interacting with poly(A) RNA and heterogeneous nuclear ribonucleoprotein K*, J Biol Chem **2002**, *277*, 40403.
[110] M. Ishaq, L. Ma, X. Wu, Y. Mu, J. Pan, J. Hu, T. Hu, Q. Fu, D. Guo, *The DEAD-box RNA helicase DDX1 interacts with RelA and enhances nuclear factor kappaB-mediated transcription*, J Cell Biochem **2009**, *106*, 296.
[111] L. Li, E. A. Monckton, R. Godbout, *A role for DEAD box 1 at DNA double-strand breaks*, Mol Cell Biol **2008**, *28*, 6413.
[112] S. Bleoo, X. Sun, M. J. Hendzel, J. M. Rowe, M. Packer, R. Godbout, *Association of human DEAD box protein DDX1 with a cleavage stimulation factor involved in 3'-end processing of pre-MRNA*, Mol Biol Cell **2001**, *12*, 3046.
[113] V. Drewett, H. Molina, A. Millar, S. Muller, F. von Hesler, P. E. Shaw, *DNA-bound transcription factor complexes analysed by mass-spectrometry: binding of novel proteins to the human c-fos SRE and related sequences*, Nucleic Acids Res **2001**, *29*, 479.

[114] J. Popow, M. Englert, S. Weitzer, A. Schleiffer, B. Mierzwa, K. Mechtler, S. Trowitzsch, C. L. Will, R. Luhrmann, D. Soll, J. Martinez, *HSPC117 is the essential subunit of a human tRNA splicing ligase complex*, Science **2011**, *331*, 760.

[115] Y. Kanai, N. Dohmae, N. Hirokawa, *Kinesin transports RNA: isolation and characterization of an RNA-transporting granule*, Neuron **2004**, *43*, 513.

[116] H. Onishi, Y. Kino, T. Morita, E. Futai, N. Sasagawa, S. Ishiura, *MBNL1 associates with YB-1 in cytoplasmic stress granules*, J Neurosci Res **2008**, *86*, 1994.

[117] C. F. Chou, W. J. Lin, C. C. Lin, C. A. Luber, R. Godbout, M. Mann, C. Y. Chen, *DEAD Box Protein DDX1 Regulates Cytoplasmic Localization of KSRP*, PLoS One **2013**, *8*, e73752.

[118] A. Perez-Gonzalez, A. Pazo, R. Navajas, S. Ciordia, A. Rodriguez-Frandsen, A. Nieto, *hCLE/C14orf166 Associates with DDX1-HSPC117-FAM98B in a Novel Transcription-Dependent Shuttling RNA-Transporting Complex*, PLoS One **2014**, *9*, e90957.

[119] J. Popow, A. Schleiffer, J. Martinez, *Diversity and roles of (t)RNA ligases*, Cell Mol Life Sci **2012**, *69*, 2657.

[120] R. Godbout, J. Squire, *Amplification of a DEAD box protein gene in retinoblastoma cell lines*, Proc Natl Acad Sci U S A **1993**, *90*, 7578.

[121] J. A. Squire, P. S. Thorner, S. Weitzman, J. D. Maggi, P. Dirks, J. Doyle, M. Hale, R. Godbout, *Co-amplification of MYCN and a DEAD box gene (DDX1) in primary neuroblastoma*, Oncogene **1995**, *10*, 1417.

[122] K. Tanaka, S. Okamoto, Y. Ishikawa, H. Tamura, T. Hara, *DDX1 is required for testicular tumorigenesis, partially through the transcriptional activation of 12p stem cell genes*, Oncogene **2009**, *28*, 2142.

[123] F. G. Barr, F. Duan, L. M. Smith, D. Gustafson, M. Pitts, S. Hammond, J. M. Gastier-Foster, *Genomic and clinical analyses of 2p24 and 12q13-q14 amplification in alveolar rhabdomyosarcoma: a report from the Children's Oncology Group*, Genes Chromosomes Cancer **2009**, *48*, 661.

[124] D. R. Germain, K. Graham, D. D. Glubrecht, J. C. Hugh, J. R. Mackey, R. Godbout, *DEAD box 1: a novel and independent prognostic marker for early recurrence in breast cancer*, Breast Cancer Res Treat **2010**.

[125] J. M. Balko, C. L. Arteaga, *Dead-box or black-box: is DDX1 a potential biomarker in breast cancer?*, Breast Cancer Res Treat **2011**.

[126] N. K. Taunk, S. Goyal, H. Wu, M. S. Moran, S. Chen, B. G. Haffty, *DEAD box 1 (DDX1) expression predicts for local control and overall survival in early stage, node-negative breast cancer*, Cancer **2011**, *118*, 888.

[127] M. Yasuda-Inoue, M. Kuroki, Y. Ariumi, *Distinct DDX DEAD-box RNA helicases cooperate to modulate the HIV-1 Rev function*, Biochem Biophys Res Commun **2013**, *434*, 803.

[128] A. Ranji, K. Boris-Lawrie, *RNA helicases: emerging roles in viral replication and the host innate response*, RNA Biol **2010**, *7*, 775.

[129] Y. Sunden, S. Semba, T. Suzuki, Y. Okada, Y. Orba, K. Nagashima, T. Umemura, H. Sawa, *Identification of DDX1 as a JC virus transcriptional control region-binding protein*, Microbiol Immunol **2007**, *51*, 327.

[130] Y. Sunden, S. Semba, T. Suzuki, Y. Okada, Y. Orba, K. Nagashima, T. Umemura, H. Sawa, *DDX1 promotes proliferation of the JC virus through transactivation of its promoter*, Microbiol Immunol **2007**, *51*, 339.

[131] L. Xu, S. Khadijah, S. Fang, L. Wang, F. P. Tay, D. X. Liu, *The cellular RNA helicase DDX1 interacts with coronavirus nonstructural protein 14 and enhances viral replication*, J Virol **2010**, *84*, 8571.

[132] P. Tingting, F. Caiyun, Y. Zhigang, Y. Pengyuan, Y. Zhenghong, *Subproteomic analysis of the cellular proteins associated with the 3' untranslated region of the hepatitis C virus genome in human liver cells*, Biochem Biophys Res Commun **2006**, *347*, 683.

[133] J. Fang, E. Acheampong, R. Dave, F. Wang, M. Mukhtar, R. J. Pomerantz, *The RNA helicase DDX1 is involved in restricted HIV-1 Rev function in human astrocytes*, Virology **2005**, *336*, 299.

[134] V. S. Yedavalli, C. Neuveut, Y. H. Chi, L. Kleiman, K. T. Jeang, *Requirement of DDX3 DEAD box RNA helicase for HIV-1 Rev-RRE export function*, Cell **2004**, *119*, 381.

References

[135] S. P. Edgcomb, A. B. Carmel, S. Naji, G. Ambrus-Aikelin, J. R. Reyes, A. C. Saphire, L. Gerace, J. R. Williamson, *DDX1 is an RNA-dependent ATPase involved in HIV-1 Rev function and virus replication*, J Mol Biol **2011**, *415*, 61.
[136] R. M. Robertson-Anderson, J. Wang, S. P. Edgcomb, A. B. Carmel, J. R. Williamson, D. P. Millar, *Single-molecule studies reveal that DEAD box protein DDX1 promotes oligomerization of HIV-1 Rev on the Rev response element*, J Mol Biol **2011**, *410*, 959.
[137] Z. Zhang, T. Kim, M. Bao, V. Facchinetti, S. Y. Jung, A. A. Ghaffari, J. Qin, G. Cheng, Y. J. Liu, *DDX1, DDX21, and DHX36 helicases form a complex with the adaptor molecule TRIF to sense dsRNA in dendritic cells*, Immunity **2011**, *34*, 866.
[138] A. Fullam, M. Schroder, *DExD/H-box RNA helicases as mediators of anti-viral innate immunity and essential host factors for viral replication*, Biochim Biophys Acta **2013**, *1829*, 854.
[139] G. J. Barton, *ALSCRIPT: a tool to format multiple sequence alignments*, Protein Eng **1993**, *6*, 37.
[140] C. D. Livingstone, G. J. Barton, *Protein sequence alignments: a strategy for the hierarchical analysis of residue conservation*, Comput Appl Biosci **1993**, *9*, 745.
[141] G. D. Van Duyne, R. F. Standaert, P. A. Karplus, S. L. Schreiber, J. Clardy, *Atomic structure of FKBP-FK506, an immunophilin-immunosuppressant complex*, Science **1991**, *252*, 839.
[142] A. D. Grossman, R. R. Burgess, W. Walter, C. A. Gross, *Mutations in the lon gene of E. coli K12 phenotypically suppress a mutation in the sigma subunit of RNA polymerase*, Cell **1983**, *32*, 151.
[143] R. G. Taylor, D. C. Walker, R. R. McInnes, *E. coli host strains significantly affect the quality of small scale plasmid DNA preparations used for sequencing*, Nucleic Acids Res **1993**, *21*, 1677.
[144] F. W. Studier, B. A. Moffatt, *Use of bacteriophage T7 RNA polymerase to direct selective high-level expression of cloned genes*, J Mol Biol **1986**, *189*, 113.
[145] J. D. Kowit, A. L. Goldberg, *Intermediate steps in the degradation of a specific abnormal protein in Escherichia coli*, J Biol Chem **1977**, *252*, 8350.
[146] D. Hanahan, *Studies on transformation of Escherichia coli with plasmids*, J Mol Biol **1983**, *166*, 557.
[147] R. K. Saiki, D. H. Gelfand, S. Stoffel, S. J. Scharf, R. Higuchi, G. T. Horn, K. B. Mullis, H. A. Erlich, *Primer-directed enzymatic amplification of DNA with a thermostable DNA polymerase*, Science **1988**, *239*, 487.
[148] Y. G. Carlo Lapid, *primerX*, bioinformatics.org/primerx/ **2003**, Bioinformatics.Org.
[149] R. H. Lambalot, C. T. Walsh, *Cloning, overproduction, and characterization of the Escherichia coli holo-acyl carrier protein synthase*, J Biol Chem **1995**, *270*, 24658.
[150] U. K. Laemmli, *Cleavage of structural proteins during the assembly of the head of bacteriophage T4*, Nature **1970**, *227*, 680.
[151] G. D. Van Duyne, R. F. Standaert, P. A. Karplus, S. L. Schreiber, J. Clardy, *Atomic structures of the human immunophilin FKBP-12 complexes with FK506 and rapamycin*, J Mol Biol **1993**, *229*, 105.
[152] H. Edelhoch, *Spectroscopic determination of tryptophan and tyrosine in proteins*, Biochemistry **1967**, *6*, 1948.
[153] C. N. Pace, F. Vajdos, L. Fee, G. Grimsley, T. Gray, *How to measure and predict the molar absorption coefficient of a protein*, Protein Sci **1995**, *4*, 2411.
[154] E. Gasteiger, A. Gattiker, C. Hoogland, I. Ivanyi, R. D. Appel, A. Bairoch, *ExPASy: The proteomics server for in-depth protein knowledge and analysis*, Nucleic Acids Res **2003**, *31*, 3784.
[155] T. Wulfmeyer, C. Polzer, G. Hiepler, K. Hamacher, R. Shoeman, D. D. Dunigan, J. L. Van Etten, M. Lolicato, A. Moroni, G. Thiel, T. Meckel, *Structural organization of DNA in chlorella viruses*, PLoS One **2012**, *7*, e30133.
[156] H. Mutschler, *Functional characterization of the chromosomally encoded toxin-antitoxin system PezAT from Streptococcus pneumoniae*, Dissertation **2011**, Dissertation 2011, 1.
[157] E. Jankowsky, M. E. Fairman, *Duplex unwinding and RNP remodeling with RNA helicases*, Methods Mol Biol **2008**, *488*, 343.
[158] M. M. Santoro, D. W. Bolen, *Unfolding free energy changes determined by the linear extrapolation method. 1. Unfolding of phenylmethanesulfonyl alpha-chymotrypsin using different denaturants*, Biochemistry **1988**, *27*, 8063.

[159] V. Consalvi, R. Chiaraluce, L. Giangiacomo, R. Scandurra, P. Christova, A. Karshikoff, S. Knapp, R. Ladenstein, *Thermal unfolding and conformational stability of the recombinant domain II of glutamate dehydrogenase from the hyperthermophile Thermotoga maritima*, Protein Eng **2000**, *13*, 501.

[160] J. Berghauser, *A reactive arginine in adenylate kinase*, Biochim Biophys Acta **1975**, *397*, 370.

[161] S. Schlee, Y. Groemping, P. Herde, R. Seidel, J. Reinstein, *The chaperone function of ClpB from Thermus thermophilus depends on allosteric interactions of its two ATP-binding sites*, J Mol Biol **2001**, *306*, 889.

[162] S. Schlee, P. Beinker, A. Akhrymuk, J. Reinstein, *A chaperone network for the resolubilization of protein aggregates: direct interaction of ClpB and DnaK*, J Mol Biol **2004**, *336*, 275.

[163] J. Reinstein, I. R. Vetter, I. Schlichting, P. Rosch, A. Wittinghofer, R. S. Goody, *Fluorescence and NMR investigations on the ligand binding properties of adenylate kinases*, Biochemistry **1990**, *29*, 7440.

[164] S. H. Thrall, J. Reinstein, B. M. Wohrl, R. S. Goody, *Evaluation of human immunodeficiency virus type 1 reverse transcriptase primer tRNA binding by fluorescence spectroscopy: specificity and comparison to primer/template binding*, Biochemistry **1996**, *35*, 4609.

[165] S. Schlee, and Reinstein, J., *Protein Folding Handbook*, Wiley-VCH Verlag GmbH & Co., Weinheim, Germany **2004**, pp.105–161, 105.

[166] W. H. Press, B. P. Flannery, S. A. Teukolsky, W. T. Vetterling, *Numerical Recipes in Pascal: the Art of Scientific Computing*, Cambridge University Press **1989**, Cambridge University Press, Cambridge.

[167] W. Kabsch, *Automatic Processing of Rotation Diffraction Data from Crystals of Initially Unknown Symmetry and Cell Constants*, J Appl Crystallogr **1993**, *26*, 795.

[168] M. D. Winn, C. C. Ballard, K. D. Cowtan, E. J. Dodson, P. Emsley, P. R. Evans, R. M. Keegan, E. B. Krissinel, A. G. Leslie, A. McCoy, S. J. McNicholas, G. N. Murshudov, N. S. Pannu, E. A. Potterton, H. R. Powell, R. J. Read, A. Vagin, K. S. Wilson, *Overview of the CCP4 suite and current developments*, Acta Crystallogr D Biol Crystallogr **2011**, *67*, 235.

[169] K. Diederichs, P. A. Karplus, *Improved R-factors for diffraction data analysis in macromolecular crystallography*, Nat Struct Biol **1997**, *4*, 269.

[170] A. J. McCoy, R. W. Grosse-Kunstleve, P. D. Adams, M. D. Winn, L. C. Storoni, R. J. Read, *Phaser crystallographic software*, J Appl Crystallogr **2007**, *40*, 658.

[171] W. Kabsch, *Xds*, Acta Crystallogr D Biol Crystallogr **2010**, *66*, 125.

[172] B. W. Matthews, *Solvent content of protein crystals*, J Mol Biol **1968**, *33*, 491.

[173] K. A. Kantardjieff, B. Rupp, *Matthews coefficient probabilities: Improved estimates for unit cell contents of proteins, DNA, and protein-nucleic acid complex crystals*, Protein Sci **2003**, *12*, 1865.

[174] N. Collaborative Computational Project, *The CCP4 suite: programs for protein crystallography*, Acta Crystallogr D Biol Crystallogr **1994**, *50*, 760.

[175] G. N. Murshudov, A. A. Vagin, E. J. Dodson, *Refinement of macromolecular structures by the maximum-likelihood method*, Acta Crystallogr D Biol Crystallogr **1997**, *53*, 240.

[176] M. D. Winn, M. N. Isupov, G. N. Murshudov, *Use of TLS parameters to model anisotropic displacements in macromolecular refinement*, Acta Crystallogr D Biol Crystallogr **2001**, *57*, 122.

[177] D. W. Buchan, S. M. Ward, A. E. Lobley, T. C. Nugent, K. Bryson, D. T. Jones, *Protein annotation and modelling servers at University College London*, Nucleic Acids Res **2010**, *38*, W563.

[178] L. H. Jensen, *Overview of refinement in macromolecular structure analysis*, Methods Enzymol **1985**, *115*, 227.

[179] A. T. Brunger, *Free R value: a novel statistical quantity for assessing the accuracy of crystal structures*, Nature **1992**, *355*, 472.

[180] V. B. Chen, W. B. Arendall, 3rd, J. J. Headd, D. A. Keedy, R. M. Immormino, G. J. Kapral, L. W. Murray, J. S. Richardson, D. C. Richardson, *MolProbity: all-atom structure validation for macromolecular crystallography*, Acta Crystallogr D Biol Crystallogr **2010**, *66*, 12.

References

[181] R. J. Laskowski RA, MacArthur MW, Kaptein R, Thornton JM, *PROCHECK - a program to check the stereochemical quality of protein structures*, <u>J App Cryst</u> **1993**, *26(Pt 2)*, 283.

[182] W. L. DeLano, *The PyMOL Molecular Graphics System*, <u>http://www.pymol.org</u> **2002**.

[183] N. A. Baker, D. Sept, S. Joseph, M. J. Holst, J. A. McCammon, *Electrostatics of nanosystems: application to microtubules and the ribosome*, <u>Proc Natl Acad Sci U S A</u> **2001**, *98*, 10037.

[184] K. Arnold, L. Bordoli, J. Kopp, T. Schwede, *The SWISS-MODEL workspace: a web-based environment for protein structure homology modelling*, <u>Bioinformatics</u> **2006**, *22*, 195.

[185] M. Hogbom, R. Collins, S. van den Berg, R. M. Jenvert, T. Karlberg, T. Kotenyova, A. Flores, G. B. Karlsson Hedestam, L. H. Schiavone, *Crystal structure of conserved domains 1 and 2 of the human DEAD-box helicase DDX3X in complex with the mononucleotide AMP*, <u>J Mol Biol</u> **2007**, *372*, 150.

[186] S. F. Altschul, W. Gish, W. Miller, E. W. Myers, D. J. Lipman, *Basic local alignment search tool*, <u>J Mol Biol</u> **1990**, *215*, 403.

[187] M. Abdelhaleem, L. Maltais, H. Wain, *The human DDX and DHX gene families of putative RNA helicases*, <u>Genomics</u> **2003**, *81*, 618.

[188] F. Sievers, A. Wilm, D. Dineen, T. J. Gibson, K. Karplus, W. Li, R. Lopez, H. McWilliam, M. Remmert, J. Soding, J. D. Thompson, D. G. Higgins, *Fast, scalable generation of high-quality protein multiple sequence alignments using Clustal Omega*, <u>Mol Syst Biol</u> **2011**, *7*, 539.

[189] T. U. Consortium, *Activities at the Universal Protein Resource (UniProt)*, <u>Nucleic Acids Res</u> **2014**, *42*, D191.

[190] H. J. Meerman, G. Georgiou, *Construction and characterization of a set of E. coli strains deficient in all known loci affecting the proteolytic stability of secreted recombinant proteins*, <u>Biotechnology (N Y)</u> **1994**, *12*, 1107.

[191] S. L. Masters, S. Yao, T. A. Willson, J. G. Zhang, K. R. Palmer, B. J. Smith, J. J. Babon, N. A. Nicola, R. S. Norton, S. E. Nicholson, *The SPRY domain of SSB-2 adopts a novel fold that presents conserved Par-4-binding residues*, <u>Nat Struct Mol Biol</u> **2006**, *13*, 77.

[192] L. Holm, P. Rosenstrom, *Dali server: conservation mapping in 3D*, <u>Nucleic Acids Res</u> **2010**, *38*, W545.

[193] J. Reed, T. A. Reed, *A set of constructed type spectra for the practical estimation of peptide secondary structure from circular dichroism*, <u>Anal Biochem</u> **1997**, *254*, 36.

[194] J. R. Lorsch, D. Herschlag, *The DEAD box protein eIF4A. 2. A cycle of nucleotide and RNA-dependent conformational changes*, <u>Biochemistry</u> **1998**, *37*, 2194.

[195] T. Hiratsuka, *New ribose-modified fluorescent analogs of adenine and guanine nucleotides available as substrates for various enzymes*, <u>Biochim Biophys Acta</u> **1983**, *742*, 496.

[196] M. A. Talavera, E. M. De La Cruz, *Equilibrium and kinetic analysis of nucleotide binding to the DEAD-box RNA helicase DbpA*, <u>Biochemistry</u> **2005**, *44*, 959.

[197] N. D. Werbeck, J. N. Kellner, T. R. Barends, J. Reinstein, *Nucleotide binding and allosteric modulation of the second AAA+ domain of ClpB probed by transient kinetic studies*, <u>Biochemistry</u> **2009**, *48*, 7240.

[198] P. Kuzmic, *DynaFit--a software package for enzymology*, <u>Methods Enzymol</u> **2009**, *467*, 247.

[199] P. Kuzmic, T. Lorenz, J. Reinstein, *Analysis of residuals from enzyme kinetic and protein folding experiments in the presence of correlated experimental noise*, <u>Anal Biochem</u> **2009**, *395*, 1.

[200] I. R. Vetter, A. Wittinghofer, *Nucleoside triphosphate-binding proteins: different scaffolds to achieve phosphoryl transfer*, <u>Q Rev Biophys</u> **1999**, *32*, 1.

[201] A. Solem, N. Zingler, A. M. Pyle, *A DEAD protein that activates intron self-splicing without unwinding RNA*, <u>Mol Cell</u> **2006**, *24*, 611.

[202] Y. Dou, T. A. Milne, A. J. Ruthenburg, S. Lee, J. W. Lee, G. L. Verdine, C. D. Allis, R. G. Roeder, *Regulation of MLL1 H3K4 methyltransferase activity by its core components*, <u>Nat Struct Mol Biol</u> **2006**, *13*, 713.

[203] N. Biris, Y. Yang, A. B. Taylor, A. Tomashevski, M. Guo, P. J. Hart, F. Diaz-Griffero, D. N. Ivanov, *Structure of the rhesus monkey TRIM5alpha PRYSPRY domain, the HIV capsid recognition module*, <u>Proc Natl Acad Sci U S A</u> **2012**, *109*, 13278.

References

[204] L. C. James, A. H. Keeble, Z. Khan, D. A. Rhodes, J. Trowsdale, Structural basis for PRYSPRY-mediated tripartite motif (TRIM) protein function, Proc Natl Acad Sci U S A **2007**, *104*, 6200.
[205] A. H. Keeble, Z. Khan, A. Forster, L. C. James, TRIM21 is an IgG receptor that is structurally, thermodynamically, and kinetically conserved, Proc Natl Acad Sci U S A **2008**, *105*, 6045.
[206] A. A. D'Cruz, N. J. Kershaw, J. J. Chiang, M. K. Wang, N. A. Nicola, J. J. Babon, M. U. Gack, S. E. Nicholson, Crystal Structure of the TRIM25 B30.2 (PRYSPRY) domain: A Key Component of Antiviral Signaling, Biochem J **2013**.
[207] H. Yang, X. Ji, G. Zhao, J. Ning, Q. Zhao, C. Aiken, A. M. Gronenborn, P. Zhang, Y. Xiong, Structural insight into HIV-1 capsid recognition by rhesus TRIM5alpha, Proc Natl Acad Sci U S A **2012**, *109*, 18372.
[208] E. Y. Park, O.-B. Kwon, B.-C. Jeong, J.-S. Yi, C. S. Lee, Y.-G. Ko, H. K. Song, Crystal structure of PRY-SPRY domain of human TRIM72, Proteins **2010**, *78*, 790.
[209] N. Biris, A. Tomashevski, A. Bhattacharya, F. Diaz-Griffero, D. N. Ivanov, Rhesus monkey TRIM5alpha SPRY domain recognizes multiple epitopes that span several capsid monomers on the surface of the HIV-1 mature viral core, J Mol Biol **2013**, *425*, 5032.
[210] Y. Yang, A. Brandariz-Nunez, T. Fricke, D. N. Ivanov, Z. Sarnak, F. Diaz-Griffero, Binding of the rhesus TRIM5alpha PRYSPRY domain to capsid is necessary but not sufficient for HIV-1 restriction, Virology **2014**, *448*, 217.
[211] D. D. Leipe, Y. I. Wolf, E. V. Koonin, L. Aravind, Classification and evolution of P-loop GTPases and related ATPases, J Mol Biol **2002**, *317*, 41.
[212] D. D. Leipe, E. V. Koonin, L. Aravind, Evolution and classification of P-loop kinases and related proteins, J Mol Biol **2003**, *333*, 781.
[213] P. Rellos, F. J. Ivins, J. E. Baxter, A. Pike, T. J. Nott, D. M. Parkinson, S. Das, S. Howell, O. Fedorov, Q. Y. Shen, A. M. Fry, S. Knapp, S. J. Smerdon, Structure and regulation of the human Nek2 centrosomal kinase, J Biol Chem **2007**, *282*, 6833.
[214] L. Aravind, Y. I. Wolf, E. V. Koonin, The ATP-cone: an evolutionarily mobile, ATP-binding regulatory domain, J Mol Microbiol Biotechnol **2000**, *2*, 191.
[215] S. Y. Moon, Y. Zheng, Rho GTPase-activating proteins in cell regulation, Trends Cell Biol **2003**, *13*, 13.
[216] J. Banroques, M. Doere, M. Dreyfus, P. Linder, N. K. Tanner, Motif III in superfamily 2 "helicases" helps convert the binding energy of ATP into a high-affinity RNA binding site in the yeast DEAD-box protein Ded1, J Mol Biol **2010**, *396*, 949.
[217] B. Schwer, T. Meszaros, RNA helicase dynamics in pre-mRNA splicing, EMBO J **2000**, *19*, 6582.
[218] A. Pause, N. Methot, N. Sonenberg, The HRIGRXXR region of the DEAD box RNA helicase eukaryotic translation initiation factor 4A is required for RNA binding and ATP hydrolysis, Mol Cell Biol **1993**, *13*, 6789.
[219] S. Rocak, B. Emery, N. K. Tanner, P. Linder, Characterization of the ATPase and unwinding activities of the yeast DEAD-box protein Has1p and the analysis of the roles of the conserved motifs, Nucleic Acids Res **2005**, *33*, 999.
[220] M. F. Perutz, Mechanisms of cooperativity and allosteric regulation in proteins, Q Rev Biophys **1989**, *22*, 139.
[221] D. J. Austin, G. R. Crabtree, S. L. Schreiber, Proximity versus allostery: the role of regulated protein dimerization in biology, Chem Biol **1994**, *1*, 131.
[222] J. R. Weir, F. Bonneau, J. Hentschel, E. Conti, Structural analysis reveals the characteristic features of Mtr4, a DExH helicase involved in nuclear RNA processing and surveillance, Proc Natl Acad Sci U S A **2010**, *107*, 12139.
[223] F. Halbach, M. Rode, E. Conti, The crystal structure of S. cerevisiae Ski2, a DExH helicase associated with the cytoplasmic functions of the exosome, RNA **2012**, *18*, 124.
[224] E. Jankowsky, RNA helicases at work: binding and rearranging, Trends Biochem Sci **2011**, *36*, 19.
[225] Z. Zhang, M. Bao, N. Lu, L. Weng, B. Yuan, Y. J. Liu, The E3 ubiquitin ligase TRIM21 negatively regulates the innate immune response to intracellular double-stranded DNA, Nat Immunol **2013**, *14*, 172.

References

[226] E. Kowalinski, T. Lunardi, A. A. McCarthy, J. Louber, J. Brunel, B. Grigorov, D. Gerlier, S. Cusack, *Structural basis for the activation of innate immune pattern-recognition receptor RIG-I by viral RNA*, Cell **2011**, *147*, 423.

[227] F. Jiang, A. Ramanathan, M. T. Miller, G. Q. Tang, M. Gale, Jr., S. S. Patel, J. Marcotrigiano, *Structural basis of RNA recognition and activation by innate immune receptor RIG-I*, Nature **2011**, *479*, 423.

[228] E. A. Gustafson, M. Yajima, C. E. Juliano, G. M. Wessel, *Post-translational regulation by gustavus contributes to selective Vasa protein accumulation in multipotent cells during embryogenesis*, Dev Biol **2011**, *349*, 440.

[229] M. J. Bennett, D. Eisenberg, *The evolving role of 3D domain swapping in proteins*, Structure **2004**, *12*, 1339.

[230] M. E. Newcomer, *Protein folding and three-dimensional domain swapping: a strained relationship?*, Curr Opin Struct Biol **2002**, *12*, 48.

[231] H. von Moeller, C. Basquin, E. Conti, *The mRNA export protein DBP5 binds RNA and the cytoplasmic nucleoporin NUP214 in a mutually exclusive manner*, Nat Struct Mol Biol **2009**, *16*, 247.

[232] C. Yan, F. Wu, R. L. Jernigan, D. Dobbs, V. Honavar, *Characterization of protein-protein interfaces*, Protein J **2008**, *27*, 59.

[233] C. Chothia, J. Janin, *Principles of protein-protein recognition*, Nature **1975**, *256*, 705.

[234] S. Jones, J. M. Thornton, *Principles of protein-protein interactions*, Proc Natl Acad Sci U S A **1996**, *93*, 13.

[235] C. O. Pabo, R. T. Sauer, *Protein-DNA recognition*, Annu Rev Biochem **1984**, *53*, 293.

[236] X. S. Liu, D. L. Brutlag, J. S. Liu, *An algorithm for finding protein-DNA binding sites with applications to chromatin-immunoprecipitation microarray experiments*, Nat Biotechnol **2002**, *20*, 835.

[237] N. M. Luscombe, R. A. Laskowski, J. M. Thornton, *Amino acid-base interactions: a three-dimensional analysis of protein-DNA interactions at an atomic level*, Nucleic Acids Res **2001**, *29*, 2860.

[238] B. M. Lunde, C. Moore, G. Varani, *RNA-binding proteins: modular design for efficient function*, Nat Rev Mol Cell Biol **2007**, *8*, 479.

[239] Y. Chen, G. Varani, *Protein families and RNA recognition*, FEBS J **2005**, *272*, 2088.

[240] S. Jones, D. T. Daley, N. M. Luscombe, H. M. Berman, J. M. Thornton, *Protein-RNA interactions: a structural analysis*, Nucleic Acids Res **2001**, *29*, 943.

[241] A. Castello, B. Fischer, K. Eichelbaum, R. Horos, B. M. Beckmann, C. Strein, N. E. Davey, D. T. Humphreys, T. Preiss, L. M. Steinmetz, J. Krijgsveld, M. W. Hentze, *Insights into RNA biology from an atlas of mammalian mRNA-binding proteins*, Cell **2012**, *149*, 1393.

[242] F. V. Karginov, J. M. Caruthers, Y. Hu, D. B. McKay, O. C. Uhlenbeck, *YxiN is a modular protein combining a DEx(D/H) core and a specific RNA-binding domain*, J Biol Chem **2005**, *280*, 35499.

[243] S. Wang, Y. Hu, M. T. Overgaard, F. V. Karginov, O. C. Uhlenbeck, D. B. McKay, *The domain of the Bacillus subtilis DEAD-box helicase YxiN that is responsible for specific binding of 23S rRNA has an RNA recognition motif fold*, RNA **2006**, *12*, 959.

[244] J. M. Caruthers, Y. Hu, D. B. McKay, *Structure of the second domain of the Bacillus subtilis DEAD-box RNA helicase YxiN*, Acta Crystallogr Sect F Struct Biol Cryst Commun **2006**, *62*, 1191.

[245] C. Maris, C. Dominguez, F. H. Allain, *The RNA recognition motif, a plastic RNA-binding platform to regulate post-transcriptional gene expression*, FEBS J **2005**, *272*, 2118.

[246] F. He, K. Saito, N. Kobayashi, T. Harada, S. Watanabe, T. Kigawa, P. Guntert, O. Ohara, A. Tanaka, S. Unzai, Y. Muto, S. Yokoyama, *Structural and functional characterization of the NHR1 domain of the Drosophila neuralized E3 ligase in the notch signaling pathway*, J Mol Biol **2009**, *393*, 478.

[247] Z. Kuang, R. S. Lewis, J. M. Curtis, Y. Zhan, B. M. Saunders, J. J. Babon, T. B. Kolesnik, A. Low, S. L. Masters, T. A. Willson, L. Kedzierski, S. Yao, E. Handman, R. S. Norton, S. E. Nicholson, *The SPRY domain-containing SOCS box protein SPSB2 targets iNOS for proteasomal degradation*, J Cell Biol **2010**, *190*, 129.

[248] C. Grutter, C. Briand, G. Capitani, P. R. Mittl, S. Papin, J. Tschopp, M. G. Grutter, *Structure of the PRYSPRY-domain: implications for autoinflammatory diseases, FEBS Lett* **2006**, *580*, 99.
[249] K. Ozato, D. M. Shin, T. H. Chang, H. C. Morse, 3rd, *TRIM family proteins and their emerging roles in innate immunity, Nat Rev Immunol* **2008**, *8*, 849.
[250] C. Weinert, C. Grutter, H. Roschitzki-Voser, P. R. Mittl, M. G. Grutter, *The crystal structure of human pyrin b30.2 domain: implications for mutations associated with familial Mediterranean fever, J Mol Biol* **2009**, *394*, 226.
[251] J. Xue, L. P. Zhu, Q. Wei, *IgG-Fc N-glycosylation at Asn297 and IgA O-glycosylation in the hinge region in health and disease, Glycoconj J* **2013**, *30*, 735.
[252] K. M. Sours, Y. Xiao, N. G. Ahn, *Extracellular-Regulated Kinase 2 Is Activated by the Enhancement of Hinge Flexibility, J Mol Biol* **2014**.
[253] R. Urich, G. Wishart, M. Kiczun, A. Richters, N. Tidten-Luksch, D. Rauh, B. Sherborne, P. G. Wyatt, R. Brenk, *De novo design of protein kinase inhibitors by in silico identification of hinge region-binding fragments, ACS Chem Biol* **2013**, *8*, 1044.
[254] V. N. Bade, J. Nickels, K. Keusekotten, G. J. Praefcke, *Covalent protein modification with ISG15 via a conserved cysteine in the hinge region, PLoS One* **2012**, *7*, e38294.
[255] B. Xiong, D. L. Burk, J. Shen, X. Luo, H. Liu, A. M. Berghuis, *The type IA topoisomerase catalytic cycle: A normal mode analysis and molecular dynamics simulation, Proteins* **2008**, *71*, 1984.
[256] D. E. Kim, H. Gu, D. Baker, *The sequences of small proteins are not extensively optimized for rapid folding by natural selection, Proc Natl Acad Sci U S A* **1998**, *95*, 4982.
[257] D. M. Karl, *Cellular nucleotide measurements and applications in microbial ecology, Microbiol Rev* **1980**, *44*, 739.
[258] I. Beis, E. A. Newsholme, *The contents of adenine nucleotides, phosphagens and some glycolytic intermediates in resting muscles from vertebrates and invertebrates, Biochem J* **1975**, *152*, 23.
[259] S. J. Ashcroft, L. C. Weerasinghe, P. J. Randle, *Interrelationship of islet metabolism, adenosine triphosphate content and insulin release, Biochem J* **1973**, *132*, 223.
[260] S. Ledoux, C. Guthrie, *Regulation of the Dbp5 ATPase cycle in mRNP remodeling at the nuclear pore: a lively new paradigm for DEAD-box proteins, Genes Dev* **2011**, *25*, 1109.
[261] J. Blumer, J. Rey, L. Dehmelt, T. Mazel, Y. W. Wu, P. Bastiaens, R. S. Goody, A. Itzen, *RabGEFs are a major determinant for specific Rab membrane targeting, J Cell Biol* **2013**, *200*, 287.
[262] R. Mattera, J. S. Bonifacino, *Ubiquitin binding and conjugation regulate the recruitment of Rabex-5 to early endosomes, EMBO J* **2008**, *27*, 2484.
[263] S. V. Paushkin, M. Patel, B. S. Furia, S. W. Peltz, C. R. Trotta, *Identification of a human endonuclease complex reveals a link between tRNA splicing and pre-mRNA 3' end formation, Cell* **2004**, *117*, 311.
[264] T. Yoshihisa, K. Yunoki-Esaki, C. Ohshima, N. Tanaka, T. Endo, *Possibility of cytoplasmic pre-tRNA splicing: the yeast tRNA splicing endonuclease mainly localizes on the mitochondria, Mol Biol Cell* **2003**, *14*, 3266.
[265] T. Yoshihisa, C. Ohshima, K. Yunoki-Esaki, T. Endo, *Cytoplasmic splicing of tRNA in Saccharomyces cerevisiae, Genes Cells* **2007**, *12*, 285.
[266] R. S. Goody, W. Hofmann-Goody, *Exchange factors, effectors, GAPs and motor proteins: common thermodynamic and kinetic principles for different functions, Eur Biophys J* **2002**, *31*, 268.
[267] R. S. Goody, *How not to do kinetics: examples involving GTPases and guanine nucleotide exchange factors, FEBS J* **2014**, *281*, 593.
[268] C. A. Tsu, O. C. Uhlenbeck, *Kinetic analysis of the RNA-dependent adenosinetriphosphatase activity of DbpA, an Escherichia coli DEAD protein specific for 23S ribosomal RNA, Biochemistry* **1998**, *37*, 16989.
[269] J. Banroques, O. Cordin, M. Doere, P. Linder, N. K. Tanner, *A conserved phenylalanine of motif IV in superfamily 2 helicases is required for cooperative, ATP-dependent binding of RNA substrates in DEAD-box proteins, Mol Cell Biol* **2008**, *28*, 3359.
[270] E. J. Tran, Y. Zhou, A. H. Corbett, S. R. Wente, *The DEAD-box protein Dbp5 controls mRNA export by triggering specific RNA:protein remodeling events, Mol Cell* **2007**, *28*, 850.

References

[271] A. R. Karow, B. Theissen, D. Klostermeier, *Authentic interdomain communication in an RNA helicase reconstituted by expressed protein ligation of two helicase domains*, FEBS J **2007**, *274*, 463.
[272] C. M. Diges, O. C. Uhlenbeck, *Escherichia coli DbpA is an RNA helicase that requires hairpin 92 of 23S rRNA*, EMBO J **2001**, *20*, 5503.
[273] M. Salazar, O. Y. Fedoroff, J. M. Miller, N. S. Ribeiro, B. R. Reid, *The DNA strand in DNA.RNA hybrid duplexes is neither B-form nor A-form in solution*, Biochemistry **1993**, *32*, 4207.
[274] S. Chakrabarti, U. Jayachandran, F. Bonneau, F. Fiorini, C. Basquin, S. Domcke, H. Le Hir, E. Conti, *Molecular mechanisms for the RNA-dependent ATPase activity of Upf1 and its regulation by Upf2*, Mol Cell **2011**, *41*, 693.
[275] A. Putnam, E. Jankowsky, *Analysis of duplex unwinding by RNA helicases using stopped-flow fluorescence spectroscopy*, Methods Enzymol **2012**, *511*, 1.
[276] T. Sengoku, O. Nureki, N. Dohmae, A. Nakamura, S. Yokoyama, *Crystallization and preliminary X-ray analysis of the helicase domains of Vasa complexed with RNA and an ATP analogue*, Acta Crystallogr D Biol Crystallogr **2004**, *60*, 320.
[277] J. C. Bischof, X. He, *Thermal stability of proteins*, Ann N Y Acad Sci **2005**, *1066*, 12.
[278] C. S. Weirich, J. P. Erzberger, J. S. Flick, J. M. Berger, J. Thorner, K. Weis, *Activation of the DExD/H-box protein Dbp5 by the nuclear-pore protein Gle1 and its coactivator InsP6 is required for mRNA export*, Nat Cell Biol **2006**, *8*, 668.
[279] C. G. Noble, H. Song, *MLN51 stimulates the RNA-helicase activity of eIF4AIII*, PLoS One **2007**, *2*, e303.
[280] A. C. Gingras, B. Raught, N. Sonenberg, *eIF4 initiation factors: effectors of mRNA recruitment to ribosomes and regulators of translation*, Annu Rev Biochem **1999**, *68*, 913.
[281] K. Berthelot, M. Muldoon, L. Rajkowitsch, J. Hughes, J. E. McCarthy, *Dynamics and processivity of 40S ribosome scanning on mRNA in yeast*, Mol Microbiol **2004**, *51*, 987.
[282] A. Dong, X. Xu, A. M. Edwards, C. Chang, M. Chruszcz, M. Cuff, M. Cymborowski, R. Di Leo, O. Egorova, E. Evdokimova, E. Filippova, J. Gu, J. Guthrie, A. Ignatchenko, A. Joachimiak, N. Klostermann, Y. Kim, Y. Korniyenko, W. Minor, Q. Que, A. Savchenko, T. Skarina, K. Tan, A. Yakunin, A. Yee, V. Yim, R. Zhang, H. Zheng, M. Akutsu, C. Arrowsmith, G. V. Avvakumov, A. Bochkarev, L. G. Dahlgren, S. Dhe-Paganon, S. Dimov, L. Dombrovski, P. Finerty, Jr., S. Flodin, A. Flores, S. Graslund, M. Hammerstrom, M. D. Herman, B. S. Hong, R. Hui, I. Johansson, Y. Liu, M. Nilsson, L. Nedyalkova, P. Nordlund, T. Nyman, J. Min, H. Ouyang, H. W. Park, C. Qi, W. Rabeh, L. Shen, Y. Shen, D. Sukumard, W. Tempel, Y. Tong, L. Tresagues, M. Vedadi, J. R. Walker, J. Weigelt, M. Welin, H. Wu, T. Xiao, H. Zeng, H. Zhu, *In situ proteolysis for protein crystallization and structure determination*, Nat Methods **2007**, *4*, 1019.
[283] A. Wernimont, A. Edwards, *In situ proteolysis to generate crystals for structure determination: an update*, PLoS One **2009**, *4*, e5094.
[284] R. Kityk, J. Kopp, I. Sinning, M. P. Mayer, *Structure and dynamics of the ATP-bound open conformation of Hsp70 chaperones*, Mol Cell **2012**, *48*, 863.
[285] C. Bieniossek, T. J. Richmond, I. Berger, *MultiBac: multigene baculovirus-based eukaryotic protein complex production*, Curr Protoc Protein Sci **2008**, Chapter 5, Unit 5 20.
[286] S. Trowitzsch, C. Bieniossek, Y. Nie, F. Garzoni, I. Berger, *New baculovirus expression tools for recombinant protein complex production*, J Struct Biol **2010**, *172*, 45.
[287] C. Okada, Y. Maegawa, M. Yao, I. Tanaka, *Crystal structure of an RtcB homolog protein (PH1602-extein protein) from Pyrococcus horikoshii reveals a novel fold*, Proteins **2006**, *63*, 1119.
[288] H. Nishi, M. Tyagi, S. Teng, B. A. Shoemaker, K. Hashimoto, E. Alexov, S. Wuchty, A. R. Panchenko, *Cancer missense mutations alter binding properties of proteins and their interaction networks*, PLoS One **2013**, *8*, e66273.
[289] E. A. Gustafson, G. M. Wessel, *DEAD-box helicases: posttranslational regulation and function*, Biochem Biophys Res Commun **2010**, *395*, 1.
[290] R. Godbout, M. Packer, S. Katyal, S. Bleoo, *Cloning and expression analysis of the chicken DEAD box gene DDX1*, Biochim Biophys Acta **2002**, *1574*, 63.

[291] R. Noguera, E. Villamon, A. Berbegall, I. Machado, F. Giner, I. Tadeo, S. Navarro, A. Llombart-Bosch, *Gain of MYCN region in a Wilms tumor-derived xenotransplanted cell line*, Diagn Mol Pathol **2010**, *19*, 33.

7. Appendix

7.1 Supplementary Figures

```
                    hhhhh  hhhhhhhhhhh     hhhhhhhhhhhhh     sssss
1HV8    1  meveymnfnelnlsdnilnairnkgfekptdiqmkviplflndeynivaq    50
           ...|:|:.:...|..|:....:..|||||.:.|||.|...  ::.
DDX1    1  ----maafscmgvmpciaqavecmdwllptdiqacsiplilggg-dvlma    45
                 ↳ hhhhhhhh         hhhhhhhhh            ssss
           translation start        Q-motif
           (Godbout et al 1994)

                 hhhhhhhhhhhhhh
1HV8   51  artgsgktasfaipliel vnen---------------------------    72
           |.||||||.:|::||:|::|.|.           SPRY domain
DDX1   46  aetgsgktgafsipviqivyetlkdqqegkkgkttiktgasvlnkwqmnp    95
           s    hhhhhhhhhhhhhhhhhh      →       ss    hhhhhhhh
                 I

1HV8   72  --------------------------------------------------    72
                                    SPRY domain
DDX1   96  ydrgsafaigsdglccqsrevkewhgcratkglmkgkhyyevschdqglc   145
                sss              ssss           sssssss      s


1HV8   72  --------------------------------------------------    72
                                    SPRY domain
DDX1  146  rvgwstmqasldlgtdkfgfgfggtgkkshnkqfdnygeeftmhdtigcy   195
              ss    sssss                                ssss


1HV8   72  --------------------------------------------------    72
                                    SPRY domain
DDX1  196  ldidkghvkfskngkdlglafeipphmknqalfpacvlknaelkfnfgee   245
              sssss


                                                      ssss  h
1HV8   73  ------------------------------------ngieaiiltpt      83
           SPRY domain                          |..:|:|:|.|:
DDX1  246  efkfppkdgfvalskapdgyivksqhsgnaqvtqtkflpnapkaliveps   295
                                                          sss       sssss


           hhhhhhhhhhhhhhh  h       ssss     hhhhhhhhhh      ssss
1HV8   84  relaiqvadeieslk---gnknlkiakiyggkaiypqikalkn-anivvg   129
           ||||.|...:.|:..|    .|..|:...|.|.|....|:..|:|  .:||||
DDX1  296  relaegtlnnikqfkkyidnpklrelliggveardqlsvlengvdivvg   345
           hhhhhhhhhhhhhhhh     sssss    Tb    hhhhhhhhh    ssss
           Ia

           hhhhhhhhhh        sssshhhhhhh   hhhhhhhhhh
1HV8  130  tpgrildhinrgtlnlknvkyfildeademlnmcfikdvekiln------   173
           ||||:..::.|.|||..|:.:|||||.:|:|.....:::.:|
DDX1  346  tpgrlddlvstgklnlsqvrflvldeadgllsqcysdfinrmhnqipqvt   395
           hhhhhhh          sssss  II          hhhhhhhhh
```

Appendix

```
            h     sssss         hh hhhhhhhhh   sssss
1HV8    174 acnkdkrillfsatmpr-eilnlakkymgdysfikak------------           209
            :..|..::::.|||:..  ::..|::|.|...::..|
DDX1    396 sdgkrlqvivcsatlhsfdvkklsekimhfptwvdlkgedsvpdtvhhvv           445
              sssss            hhhhhhhhh      sssss          sssss
                   III
                                       Domain 1    Linker

                        sssss hhhhh                hhhhhhhhh
1HV8    210 --in------------anieqsyvevnener---------fealcrllk-          235
            :|              ::|....|...:|.|         .||:  ::||
DDX1    446 vpvnpktdrlwerlgkshirtddvhakdntrpganspenwseai-kilkg            494
            ss     hhhhhh                            hhhhhh  hhhhh
                              Linker      Domain 2

                    ssss     hhhhhhhhhhhhhhh                sss
1HV8    236 ---------nkefyglvfcktkrdtkelasmlrdig---------fkaga           267
            :|....::||:||.|....|                         |....
DDX1    495 eyavraikehkmdqaiifcrtkidcdnleqyfiqqgggpdkkghqfscvc            544
            hhhhhhhh       sssss        hhhhhhhhhh                sssss
                            IV

                s    hhhhhhhhhhhhhh    ssss hhhhhh           ssss
1HV8    268 ihgdlsqsqrekvirlfkqkkiriliatdvmsrgidvndlncvinyhlpq            317
            :|||....:|::.:..||:..:|.||.|||.:|||||::.:..|||..||.
DDX1    545 lhgdrkpherkqnlerfkkgdvrflictdvaargidihgvpyvinvtlpd            594
             s   hhhhhhhhhhh       ssss                  sssss
                                          V

            hhhhhhhhh        ssssss    hhhhhhhh  hhhhh
1HV8    318 npesymhrigrtgragkkgkaisiinrreykklryi--eramklkikklk            365
            ..::|:|||||.|||.:.|.|||::...:.|....::  .|........:||
DDX1    595 ekqnyvhrigrvgraermglaislvatekekvwyhvcssrgkgcyntrlk            644
            ss                   sssss      hhhhhhhhhhhh
              VI


1HV8    366 fg------------------------------------------------           367
            ..
DDX1    645 edggctiwynemqllseieehlnctisqvepdikvpvdefdgkvtygqkr            694
            Hhhhhhhhhhhhhhh                  |       sss     |
                                             674              694


1HV8    367 -------------------------------------------------           367
DDX1    695 aagggsykghvdilaptvqelaalekeaqtsflhlgylpnqlfrtf               740
            hhhhhhhhhhhhhhhhhhhhhhhhhhh
```

Figure 7.1.1 **Sequence alignment of human DDX1 with MjDEAD (1HV8).** The complete sequence of human DDX1 (aa 1-740) was aligned with the sequence of a DEAD-box protein from the hyperthermophile *Methanococcus jannaschii*. The structure of the *M. jannaschii* protein was solved and deposited in the PDB as 1HV8 [33]. Conserved sequence motifs and the domain boundaries are indicated in blue. The large SPRY-domain insertion between Walker A and B motifs is indicated in brown. Secondary structure is indicated by h=α-helix and s=β-strand. For DDX1 secondary structure was predicted using psipred [177]. Pairwise sequence alignment was done using needle (http://www.ebi.ac.uk/Tools/psa/emboss_needle).

Appendix

```
DDX1    MAAFSEMGVMPEIAQAVEEMDWLLPTDIQAESIPLILGGGDVLMAAETGSGKTGAFSIPV  60
Dbp2p   ---------------------------------MTYGGRDQQY---------------   10
                                         :  **  *

DDX1    IQIVYETLKDQQEGKKGKTTIKTGASVLNKWQMNPYDRGSAFAIGSDGLCCQSREVKEWH 120
Dbp2p   ---------------------------NKTNYKSRGGDFRGGRNS------------  28
                                   :  *  .**.  *   * :.

DDX1    GCRATKGLMKGKHYYEVSCHDQGLCRVGWSTMQASLDLGTDKFGFGFGGTGKKSHNK--Q 178
Dbp2p   ----------DRNSYN----------------D--RPQGGNYRGGFGGGRSNYNQPQELI  59
                  :. *:                  ::  ****  .:  .: :

DDX1    FDNYGEEFTMHDTIGC--YLDI-------DKGHVKFSKNGKD--LGLAFEIPPHMKNQAL 227
Dbp2p   KPNWDEELPKLPTFEKNFYVEHESVRDRSDSEIAQFRKENEMTISGHDIPKPITTFDEAG 119
         *;  **;      *;    *::         *.   ..:* *: :    *   :    *    ::*

DDX1    FPACVLKNAELKFNFGEEEFKFPPKD---G---------FVALSKAPDGYIVKSQHSGNA 275
Dbp2p   FPDYVLNEVKA------EGFDKPTGIQCQGWPMALSGRDMVGIAATGSGKTLSYCLPGIV 173
         **  **:: .:        * *. *          *.:: :  .*  :.       *.

DDX1    QVTQTKF--LPNAPKALIVEPSRELAEQTLNWIKQFKKYIDNPKLRELLIIGGVAARDQL 333
Dbp2p   HINAQPLLAPGDGPIVLVLAPTRELAVQIQTE---CSKFGHSSRIRNTCVYGGVPKSQQI 230
         ::.    :     :.*  .*:: *;****  *   .:       .*:  ..  :*:    *** :*:

DDX1    SVLENGVDIVVGTPGRLDDLVSTGKLNLSQVRFLVLDEADGLLSQGYSDFINRMHNQIPQ 393
Dbp2p   RDLSRGSEIVIATPGRLIDMLEIGKTNLKRVTYLVLDEADRMLDMGFEPQIRKIVDQIRP 290
         *..*  :**:.***** *::.  ** **.:*  *:****** *.    * . *.:: :**

DDX1    VTSDGKRLQVIVCSATLHSFDVKKLSEKIMHFPTUWDLKGEDSVPDTVHHVVVPVNPKTD 453
Dbp2p   DR---Q---TLMWSAT-WPKEVKQLAADYLNDPIQVQVGSLE----------------  325
         :      .:: ***    :**:*: . . *  *::. :*

DDX1    RLWERLGKSHIRTDDVHAKDNTRPGANSPEMWSEAIKILKGEYAVRAIKEHKMDQAIIFC 513
Dbp2p   ------LSASHNITQIVEVVSD----------FEKRD--RLNKYL-ETASQDNEYKTLIFA 367
              *.  **   *:  *..  .:           :.:           :*    .:  .:.:    :::**.

DDX1    RTKIDCDNLEQYFIQQGGGPDKKGHQFSCVCLHGDRKPHERKQNLERFKKGDVRFLICTD 573
Dbp2p   STKRMCDDITKYLRED---------GWPALAIHGDKDQRERDWVLQEFRNGRSPIMVATD 418
         **  **::  :*:  ::          :  .:;***:. :**.  *:.*:.*   :::.**

DDX1    VAARGIDIHGVPYVINVTLPDEKQNYVHRIGRVGRAERMGLAISLVATEKEKVWYHVCSS 633
Dbp2p   VAARGIDVKGINYVINYDMPGNIEDYVHRIGRTGRAGATGTAISFFTEQMKGLGAKLISI 478
         *******:.*:. ****  :*  : :::*******.***  * ***:.:  :::  : ::    *

DDX1    RGKGCYNTRLKEDGGCTIWYNEMQLLSEIEEHLNCTISQVEPDI-KVPVDEF-------- 684
Dbp2p   ------------------------MREANQN-------IPPELLKYDRRSYGGGHPRYG 506
                                  :  *: ::.         *: :*    .:

DDX1    ---DGKWTYGQKRAAGG--GSYKGHVDILAPTVQELAALEKEAQTSFLHLGYLPNQLFRT 739
Dbp2p   GGRGGRGGYGRRGGYGGGRGGYGGNRQ-------RDGGWGNRGRSNY------------  546
         *:  **::.    **  *.** *. :         ..  :..::.:
 
DDX1    F      740
Dbp2p   -      546
```

Figure 7.1.2 **Sequence alignment of human DDX1 with yeast Dbp2.** Human DDX1 is aligned with the closest yeast homologue, the DEAD-box protein Dbp2p. The Walker A motif (sequence GSGKT) is highlighted in both sequences to show disparities in the alignment of the C-terminal RecA-like domain.

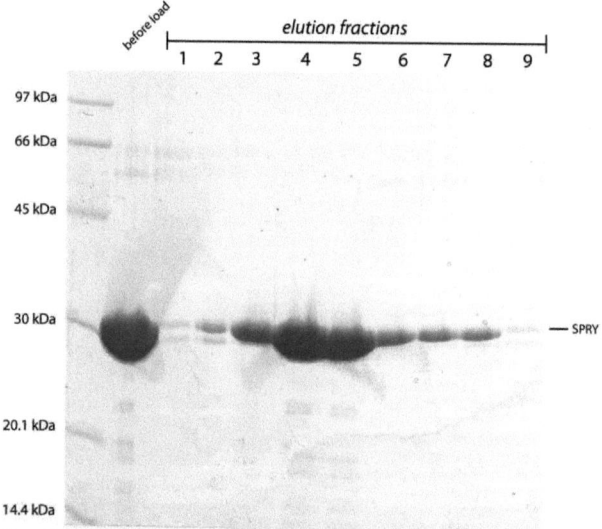

Figure 7.1.3 **Purification of the separated SPRY domain.** A construct of the separated SPRY domain of DDX1 (amino acids 72-283) was purified by initial Ni^{2+}, followed by heparin affinity chromatography. Elution fractions from the heparin column were loaded on a 15% SDS-gel.

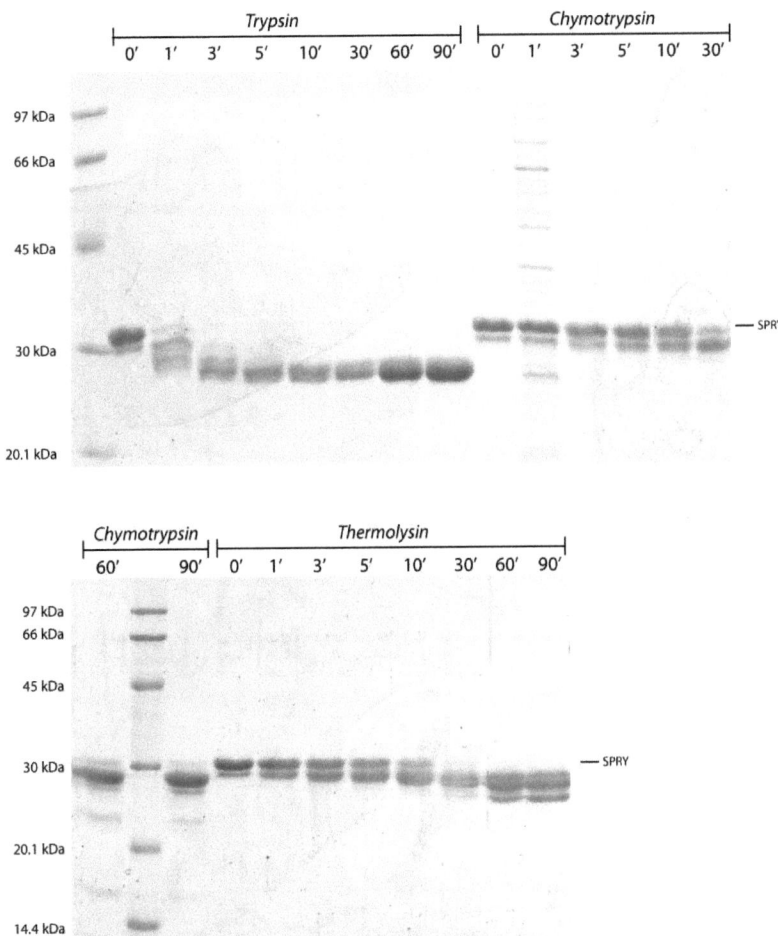

Figure 7.1.4 **Limited proteolysis of the SPRY domain with Trypsin, Chymotrypsin and Thermolysin.** A Construct of the SPRY domain of DDX1 (amino acids 72-283) was digested with different proteases (1 µg protease per 1600 µg protein) and reactions were quenched at the indicated time points (in min). Reactions were separated on a 15% SDS-gel and degradation bands were identified by MALDI-MS.

Appendix

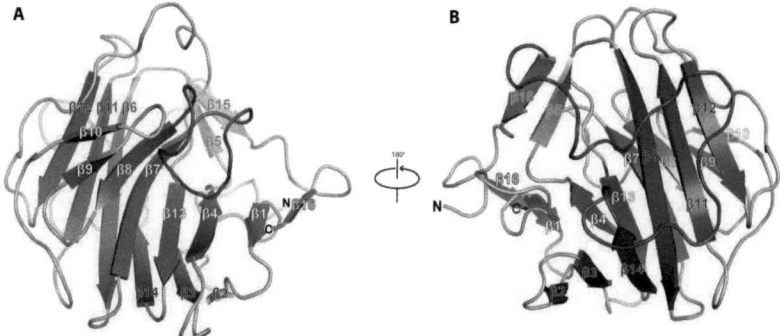

Figure 7.1.5 **Long loops cover hydrophobic patches on the surface of the SPRY domain.**, On both sides of the SPRY β-sandwich fold extended loops are found that clamp the compact domain. **A**, on the concave side, loop D (depicted in purple) covers a hydrophobic patch. **B**, on the convex side the loop between β14 and β15 (depicted in purple) covers a hydrophobic patch of β-sheet 2. A ribbon representation of the β-sandwich fold of the SPRY domain is shown. β-sheet 1 is depicted in blue, β-sheet 2 in red and β-sheet 3 in green. The different strands and N- and C-termini are labeled.

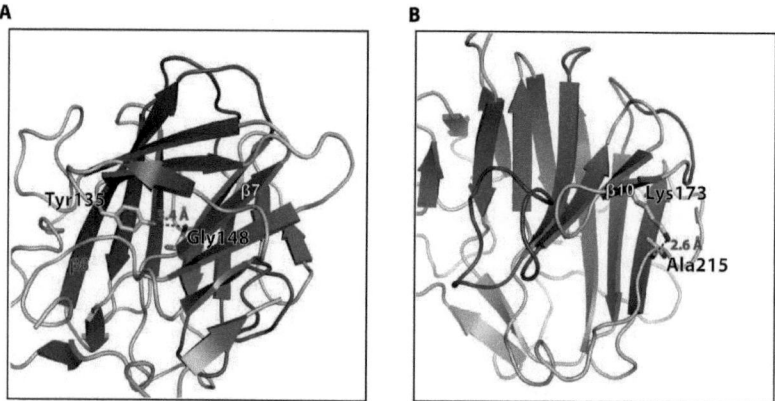

Figure 7.1.6 **Hydrogen bonds stabilize the β-sandwich fold.** The SPRY domain is depicted in ribbon representation with β-sheet 1 in blue, β-sheet 2 in red and β-sheet 3 in green. **A**, hydrogen bond between Tyrosine 135 of strand β6 and Glycine 148 of strand β7 (3.4 Å) that holds both sheet together is shown. **B**, hydrogen bond between Lysine 173 of strand β10 and Alanine 215 of the loop connecting strand β12 and strand β13 (2.6 Å) that stabilizes the fold is shown.

Appendix

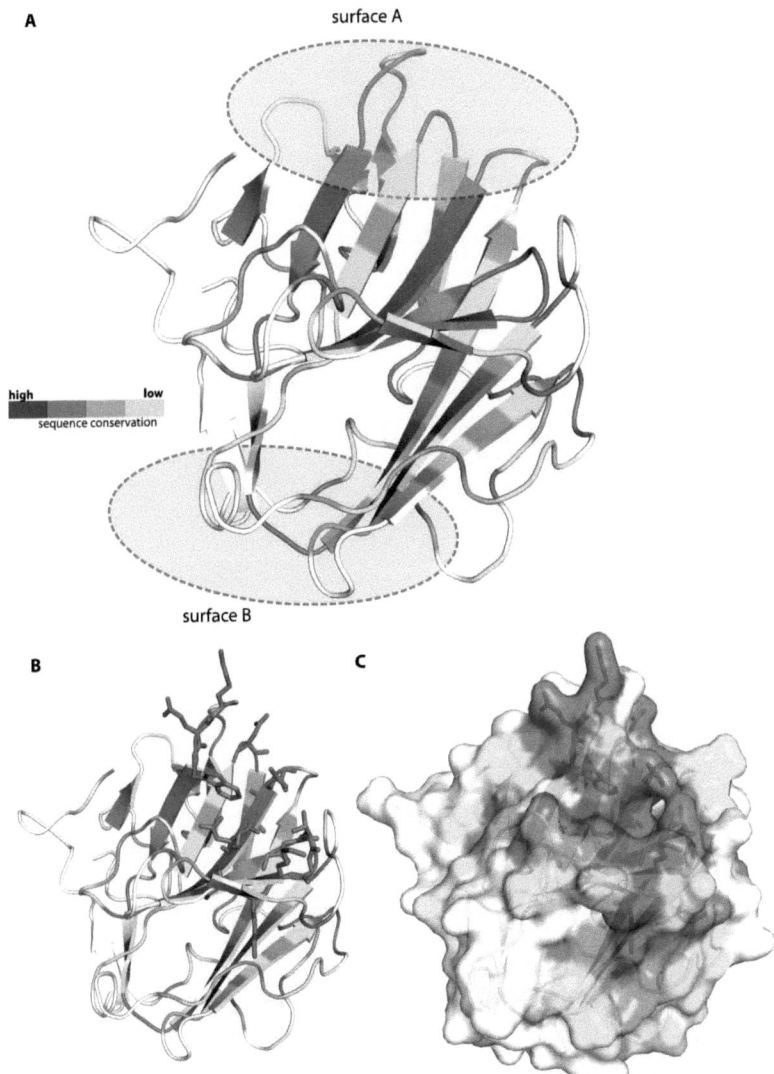

Figure 7.1.7 **Sequence conservation of the DDX1-SPRY domain amongst model organisms.** A few residues on the surface form a conserved patch that is distinct from surface A. **A**, the sequence conservation between different eukaryotic model organisms was mapped on the structure of the human DDX1-SPRY domain. Surfaces A and B, formed by the loops that connect the β-strands are indicated by grey shading. **B**, residues that form the conserved surface patch, designated surface X, are shown as sticks. **C**, a surface representation of the sequence conservation is shown with the conserved residues that from surface X shown as sticks.

Appendix

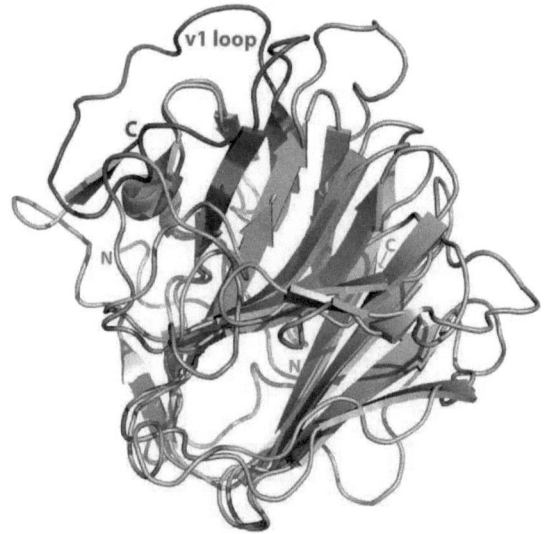

Figure 7.1.8 **Comparison of DDX1 SPRY with the distantly related SPRY domain of Trim21.** The superposition of DDX1 SPRY with Trim21 SPRY(PDB entry 3ZO0/2VOL[205])[205] as obtained from the DALI server[192] is shown. Both proteins are depicted in ribbon representation. DDX1 SPRY is depicted in green and Trim21 SPRY is depicted in blue. The region of highest divergence, the loop connecting β-strands 3 and 4 (referring to DDX1) is indicated by more intense color shading and labeled v1 loop. DALI superposition resulted in a Z-Score of 14.6 and an overall rmsd of 2.1 Å for the alignment of 138 residues[192]

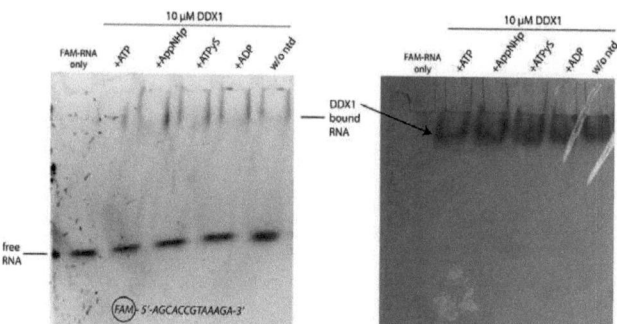

Figure 7.1.9 **Binding of DDX1 to 5'-labeled RNA observed via gel-shift.** Binding of a 13-mer RNA that was labeled at the 5'-end with a FAM-label to DDX1 was observed via a band-shift in a native gel. 10 µM of DDX1-728 were incubated with labeled 13-mer RNA of indicated sequence and 10 mM of different nucleotides (as indicated) for 20 min at 277 K. Reactions were separated on a 5%/15% native acryl-amide-gel with CHES-NaOH pH 10 as running buffer. The FAM label was visualized by illumination at 302 nm. Protein was stained with Instant-blue to confirm co-migration with the RNA signal (right side).

Appendix

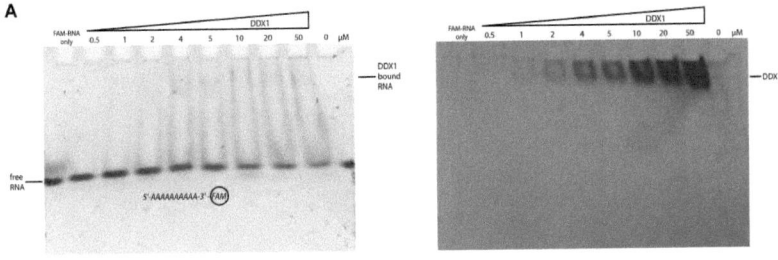

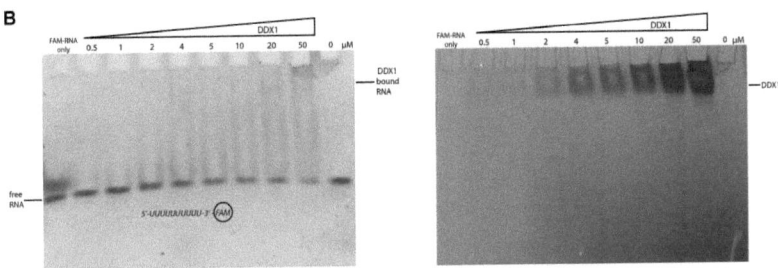

Figure 7.1.10 **Binding of DDX1 to poly A or poly U sequences observed via gel-shift.** Binding of a fluorescent FAM-labeled RNA (labeled at the 3'-end) to DDX1 was observed via a band-shift in a native gel. Increasing concentrations of DDX1-728 were incubated with labeled 10-mer RNA of indicated sequence for 20 min at 277 K. Reactions were separated on a 5%/15% native acryl-amide-gel with CHES pH 10 as running buffer. **A**, binding of DDX1 to a 10mer poly-A sequence is shown. **B**, binding of DDX1 to a 10mer poly-U sequence is shown.
The FAM label was visualized by illumination at 302 nm. Protein was stained with Coomassie-brilliant-blue to confirm positioning at the same height as the RNA signal (see right side).

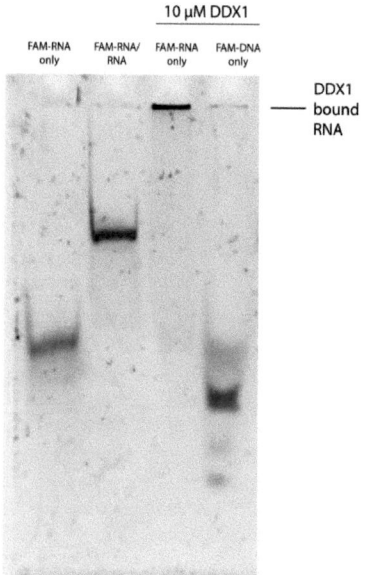

Figure 7.1.11 **Binding of DDX1 to RNA only.** Binding of DDX1 to the same sequence of either fluorescent FAM-labeled RNA or DNA (labeled at the 3'-end) was observed on a native 15 % (w/v) TBE (pH 8.3) helicase gel. 10 µM of DDX1-728 were incubated with labeled 13-mer oligonucleotide[49] (either RNA or DNA, sequence as in figure 7.1.9) for 60 min at 277 K. Reactions were separated on a 15 % (w/v) native acryl-amide-gel with 0.5 x TBE pH 8.3 as running buffer. The FAM label was visualized by illumination at 302 nm.

Appendix

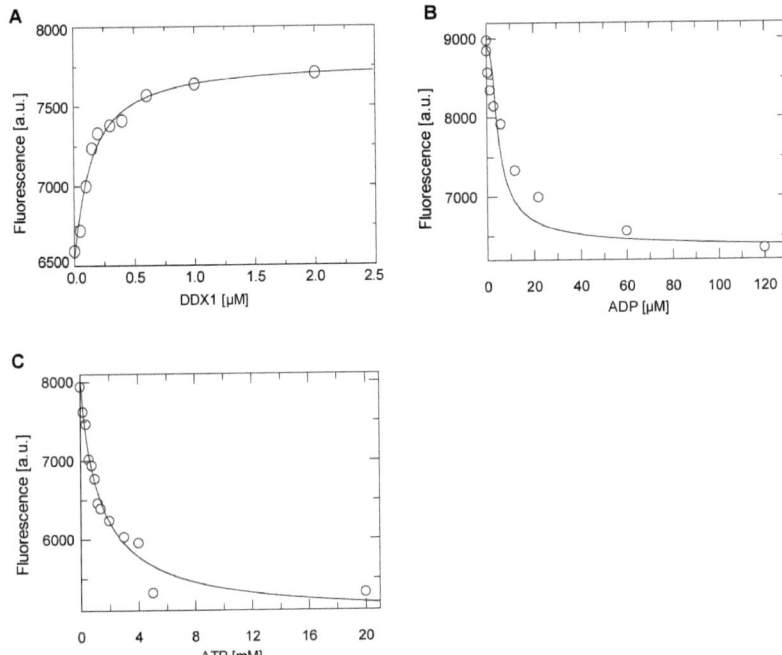

Figure 7.1.12 **DDX1 equilibrium titrations with mant-dADP.** **A**, to obtain a $K_{d,mantdADP}$ 0.05 μM mant-dADP were titrated with DDX1 protein. **B**, to obtain an ADP-affinity ($K_{d,ADP}$) a mixture of 0.2 μM mant-dADP and 3 μM DDX1 was titrated with unlabeled ADP. **C**, a mixture of 0.2 μM mant-dADP and 1 μM DDX1 was titrated with ATP to obtain a $K_{d,ATP}$. Titration was conducted in the presence of an ATP-regenerating system.
Binding data in A were fitted with the quadratic equation and the resulting affinity constant was used as input parameter for the competition experiments in B and C that were fitted with the cubic equation. All parameters obtained from these experiments are shown in table 7.1.

Appendix

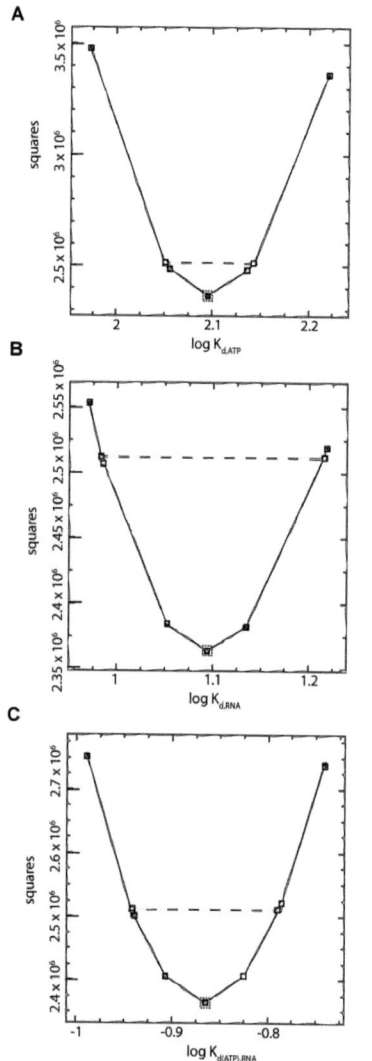

Figure 7.1.13 **Confidence intervals of parameters from global fit**. All equilibrium titration experiments were fitted globally by numeric iteration in the program Dynafit[198]. The affinity constants obtained by the global fit are shown in table 3.4. **A**, confidence interval for the optimization of the kinetic parameter for ATP binding, $K_{d,ATP}$ is shown. Best fit values for the parameter are plotted against the sum-of-squares of the deviation to the actual data. The final value that was reported is marked with a dashed square. **B**, confidence interval for the optimization of the kinetic parameter for RNA binding, $K_{d,RNA}$ is shown. Representation as in A **C**, confidence interval for the optimization of the kinetic parameter for RNA binding to ATP-saturated DDX1, $K_{d(ATP),RNA}$ is shown. Representation as in A.

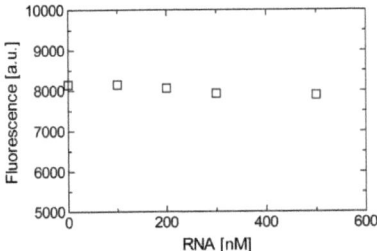

Figure 7.1.14 **Effect of RNA titration on ADP binding.** **A**, to test a potential influence of ADP on the RNA affinity of DDX1 a titration experiment was performed. A complex of 0.2 µM mant-dADP and 1 µM DDX1 was partially displaced by addition of 3 µM ADP. In a next step the actual experiment (as shown in the plot) was started by titrating 10mer polyA RNA. No RNA stimulated ADP binding (which would lead to a significant decrease in fluorescence) could be observed.

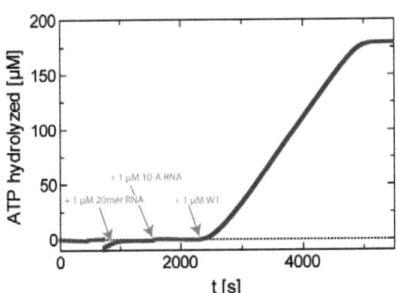

Figure 7.1.15 **DDX1 ATPase activity of K52A variant.** A variant of DDX1, where the canonical Walker A lysine had been replaced with an alanine was tested for its ATP hydrolysis activity in a coupled assay. 1 µM DDX1 K52A was used and during the time coure of the experiment 1 µM 20mer RNA (see mat&meth), 1 µM 10mer polyA RNA and 1 µM DDX1 WT were added as indicated. For the assay 4 U LDH and 2.8 U PK were mixed with 1 mM ATP in 50 mM Tris buffer pH 8.0, containing 0.5 mM NADH, 0.8 mM PEP, 250 mM NaCl, 1 mM EDTA and 15 mM KCl. Reactions were started by addition of 5 mM MgCl$_2$ and 1 µM of DDX1 K52A. Decrease in A$_{340}$ is monitored and serves as a measure for the consumption of ATP (=generation of ADP). The K52A variant of DDX1 was ATPase deficient, which shows that the activity observed for WT DDX1 is protein specific and not cause by contaminations.

Appendix

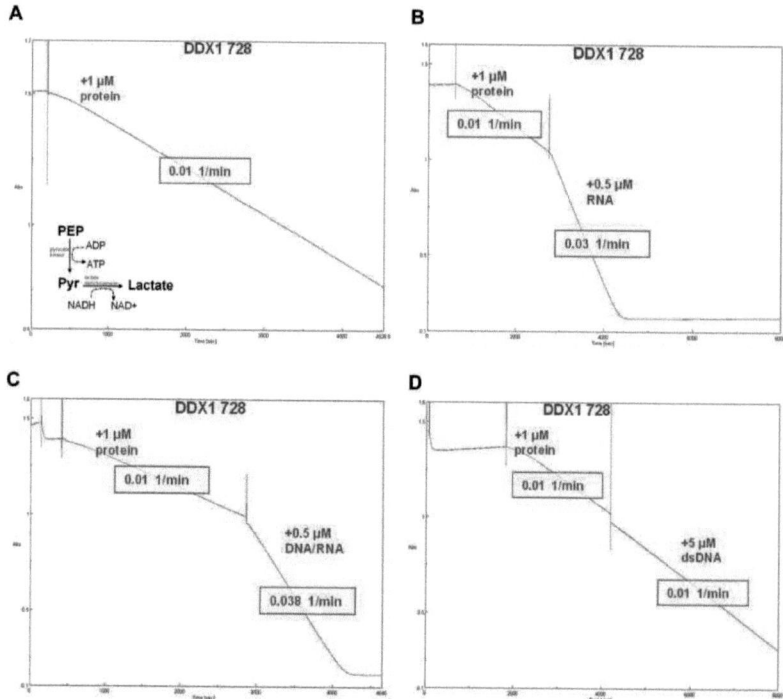

Figure 7.1.16 DDX1 ATPas stimulation by RNA and DNA. DDX1 (construct DDX1-728) exhibits an intrinsic ATPase activity that can be stimulated by single stranded RNA or DNA/RNA hybrids, but not by dsDNA. For the assay 4 U LDH and 2.8 U PK were mixed with 1 mM ATP in 50 mM Tris buffer pH 8.0, containing 0.5 mM NADH, 0.8 mM PEP, 250 mM NaCl, 1 mM EDTA and 15 mM KCl. Reactions were started by addition of 5 mM $MgCl_2$ and the indicated concentrations of protein. Decrease in A_{340} is monitored and serves as a measure for the consumption of ATP (=generation of ADP).
A, intrinsic ATPase activity of 1 µM DDX1-728 with a rate of 0.01 ΔAbs/min, which equals 1.61 µM (NADH)/min. **B**, RNA-stimulated ATPase rate of 0.03 ΔAbs/min equals 4.82 µM (NADH)/min. RNA that was used to stimulate was a 29-mer according to [111]. Note that the stimulation of the hydrolysis rate of 1 µM DDX1 by 0.5 µM RNA did not differ between different oligonucleotide sequences that were tested. All short RNA did stimulate hydrolysis activity by a factor of ~ 3. **C**, ATP-hydrolysis is stimulated by a DNA/RNA hybrid, which gives an ATPase rate of 0.038 ΔAbs/min that equals 6.13 µM(NADH)/min. DNA RNA hybrid corresponds to 41mer DNA annealed to a 29-mer RNA according to [94] **D**, double-stranded DNA does not stimulate the ATPase rate of DDX1.

Appendix

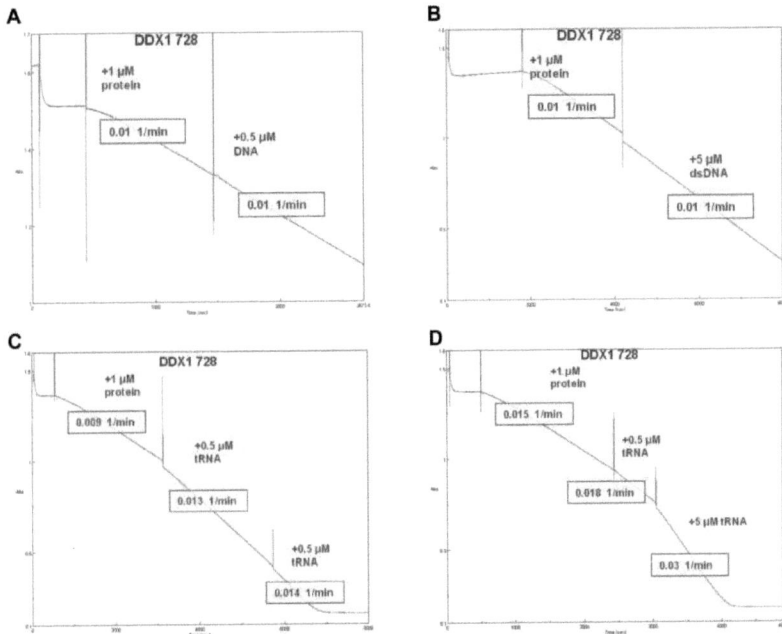

Figure 7.1.17 **DDX1 ATPase stimulation by tRNA.** DDX1 exhibits an intrinsic ATPase activity that it not stimulated by DNA, but can be stimulated by high concentrations of tRNA. **A**, intrinsic ATPase activity of 1 µM DDX1-728 does not get stimulated by addition of DNA **B**, intrinsic ATPase activity does not get stimulated by addition of dsDNA **C**, highly structured tRNA-Phe does barely stimulate ATPase activity **D**, only at high concentrations of tRNA-Phe, a stimulating effect on ATPase activity is seen.

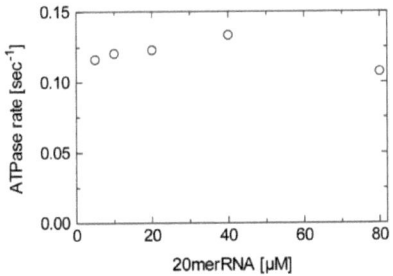

Figure 7.1.18 **ATPase activity of DDX1.** RNA concentration dependence of the ATPase rate at saturating ATP concentrations and high concentrations of RNA is shown. 1 µM DDX1 was incubated with indicated concentrations of a 20mer RNA in the presence of 10 mM ATP-MgCl$_2$. The measured maximal initial reaction velocity (= ATPase rate) did not differ from measurements with 10 mM ATP-MgCl$_2$ in the absence of RNA.

Appendix

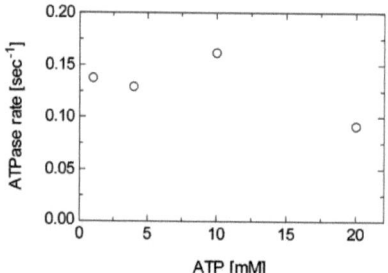

Figure 7.1.19 **ATPase activity of DDX1.** ATP concentration dependence of the ATPase rate at saturating RNA concentrations is shown. 1 µM DDX1 was incubated with indicated concentrations of ATP in the presence of 40 µM 20mer RNA. The measured maximal initial reaction velocity (= ATPase rate) did not differ in measurements with different ATP concentrations.

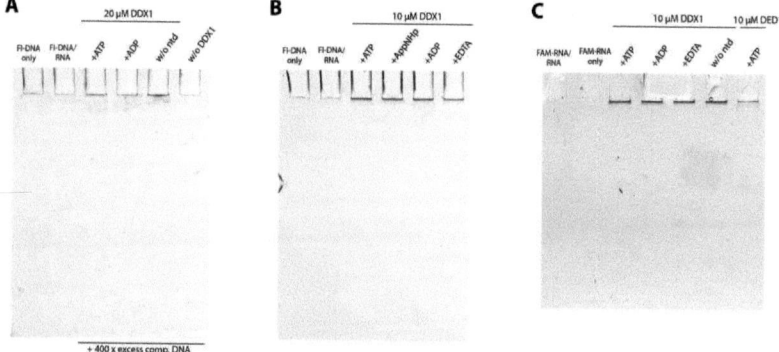

Figure 7.1.20 **DDX1 helicase gels, stained with Coomassie.** The 15 % (w/v) native TBE-gels (pH 8.3) that were used for the DDX1 helicase assay (see section 3.3.9, figures 3.4.1 and 3.4.2) were stained with InstantBlue® Coomassie stain to visualize the protein. In contrast to the EMSA gels (see section 3.3.3), in this case the protein did not enter the TBE-gels, but was stuck in the gel pockets. **A**, InstantBlue® Coomassie stain of the helicase gel, depicted in figure 3.4.1 a. **B**, InstantBlue® Coomassie stain of the helicase gel, depicted in figure 3.4.1 b. **C**, InstantBlue® Coomassie stain of the helicase gel, depicted in figure 3.4.2.

Appendix

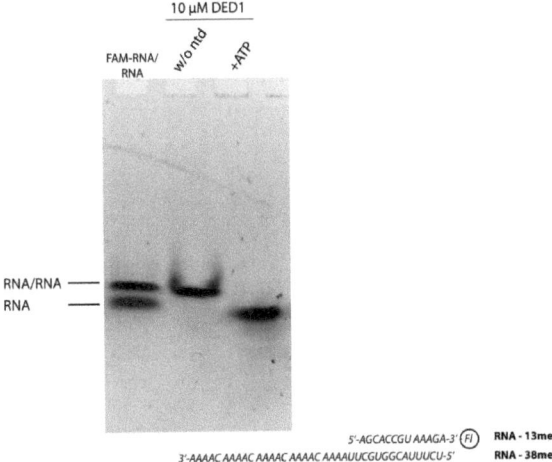

Figure 7.1.21 **DED1 helicase assay with dsRNA substrate.** A solution of 10 µM DED1 was incubated with 10 mM nucleotides and 0.5 µM double-stranded RNA substrate as indicated. Reactions were incubated at 310 K for 60 min and then separated on a 15 % TBE-gel. The dsRNA substrate was designed according to Jankowsky, 2008[157].

7.2 Supplementary tables

Table 7.1 **mant-dADP dependent equilibrium binding constants**

mant-dADP binding			
$K_{d,mantdADP}$	0.1237	+/- 0.028	µM

ADP binding (mant-dADP)			
$K_{d,ADP}$	0.0963	+/- 0.0038	µM

ATP binding (mant-dADP)			
$K_{d,ATP}$	129.3	+/- 25.8	µM

Table 7.2. **nucleotide binding affinities of DDX1 obtained in the global fit** (by numeric iteration)

mant-ADP binding				Δanalytical fit
$K_{d,mantADP}$	0.5036	± 0.2559	µM	0.3796 µM

ATP binding				
$K_{d,ATP}$	124.927	± 21.088	µM	4.373 µM
$K_{d,(RNA),ATP}$	1.4	± 21.088	µM	3.64 µM

Table 7.3 **Truncated constructs of DDX1 used in crystallization screens.**

Construct	Co-factor/substrate	Screens
DDX1-432	apo	JCSG suite
DDX1-432	AppNHp	JCSG suite
DDX1-610	apo	Classics
DDX1-610	ATPγS	Classics, JCSG suite
DDX1-610	AppNHp	Classics, JCSG suite
DDX1-655	AppNHp	Classics
DDX1-694woSPRY	apo	JCGS suite, PEG suite

-> constructs listed in the table did not form X-ray diffracting protein crystals with the screening conditions tested

7.3 Supplementary results on construct design, characterization and crystallization

7.3.1 C-terminally truncated DDX1 constructs

Based on analysis of the degradation products of full-length DDX1 by MALDI-MS combined with secondary structure predictions, several C-terminally truncated protein variants were generated to map for stable domain boundaries. Out of the generated fragments, constructs DDX1-655, DDX1-648 and DDX1-610 (see domain boundaries in **Figure 3.1**) were found in the soluble fraction upon bacterial expression. These three C-terminally truncated variants of human DDX1 could be purified to higher homogeneity than all other constructs.

All three constructs were used in crystallization experiments and activity assays (see **Table 7.3**). Unfortunately they did not yield protein crystals (see section on crystallization below) and were ATPase deficient (data not shown). This functional defect was probably caused by too stringent truncation as in the case of DDX1-610 already signature motif VI was missing. Due to the mentioned problems the constructs were abandoned and not used for further characterization of DDX1.

7.3.2 RecA-like domain 1 of DDX1

DEAD-box proteins consist of two RecA-like helicase domains that are connected by a flexible linker and can therefore sample a variety of different orientations of the two domains relative to each other. This flexibility interferes with crystallization. To be able to study the arrangement of the unique SPRY domain of DDX1 in the context of RecA-like domain 1 (where the SPRY domain forms an insertion) a modular approach was undertaken. The separated RecA-like domain 1 of DDX1 including the SPRY domain (amino acids 1-432, see **Figure 3.1**) was cloned. This construct, designated DDX1-432 could be expressed as a soluble protein in *E. coli* and purified to high homogeneity. The construct was stable and used in crystallization trials (see **Table 7.3**).

Appendix

7.3.3 Co-crystallization of C-terminally truncated DDX1 in complex with non-hydrolyzable ATP-analogs

Purifications protocols were established for the generation of several DDX1 constructs that almost represent the full-length protein, only bearing slight C-terminal truncations (**see section 7.3.1**). The constructs DDX1-674, DDX1-694 and DDX1-728 were used in crystallization experiments. Out of the three constructs, DDX1-674 could be purified to highest homogeneity as assessed by SDS-gels (**Figure 7.3.1 b**).

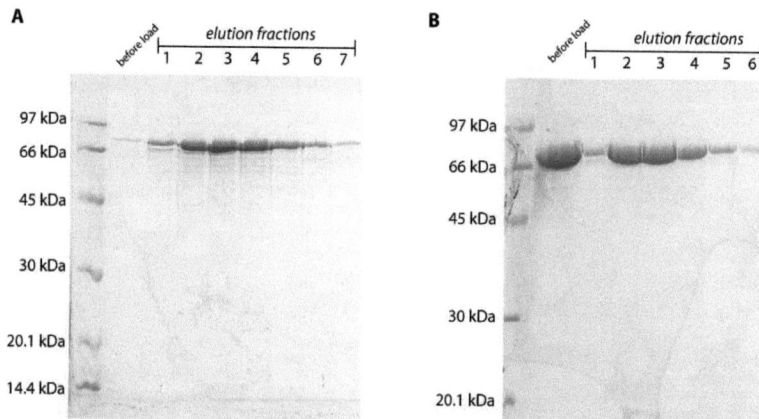

Figure 7.3.1 **Size exclusion chromatography run with DDX1-694/674.** Fractions of S6 size-exclusion chromatography runs were separated on a 15% SDS-gel and identity of bands checked by MALDI-MS. **A**, protein construct DDX1-694 that is C-terminally truncated at Arg 694 was purified by a combination of 4 chromatographic steps. **B**, construct DDX1-674 was purified to high homogeneity by a combination of 4 chromatographic steps.

DLS measurements with protein DDX1-674 (**Figure 7.3.2 b**) showed a peak width of the distribution of hydrodynamic radii of only 24.4 % RSD. Using temperature-dependent CD-unfolding a melting point of 323.6 +/- 0.08 K was determined (**Figure 7.3.2 a**), indicating a reasonable stable protein, suitable for structural studies.

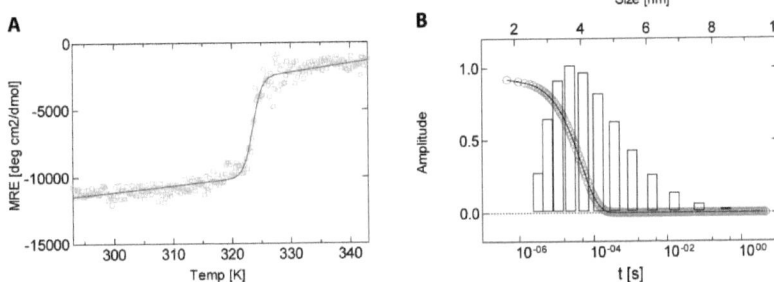

Figure 7.3.2 **Stability and homogeneity of DDX1-674 construct.** A CD melting curve of DDX1-674 monitored at 222 nm and fitted to a two-state-unfolding process yields a Tm of 323.6 +/- 0.08 K. CD values are in mean residue ellipticity = MRE. **B**, Data from DLS measurement with DDX1-674. The combined autocorrelation function (grey circles) is shown on the lower x-axis and a fit to this data is depicted as a black line.. The upper x-axis shows the distribution of the hydrodynamic radii by relative mass (amplitude of each bar indicates the percentage of the total mass of the sample) as obtained from the fit. The fit did yield a peak in the mass-distribution of hydrodynamic radii at 4.15 nm (corresponding to a molecular weight of 95.38 kDa) and a peak width of 24.4 % RSD, which indicates a high degree of sample homogeneity.

Crystallization trials were set-up using commercially available screens (see **methods section 2.5.1**). DDX1-674 at 20 mg/ml (= 250 µM) was screened in apo and in complex with 10 mM ATPγS and AppNHp, however, no crystal formation was observed. In contrast, a variant of DDX1-674, where the N-terminal His-Tag had been cleaved off during purification (=DDX1-674ΔHis-Tag), showed crystal formation in the presence of an eightfold molar excess of ATPγS. Crystals were small and heavily intertwined. Crystal formation was observed with reservoir solutions containing either 10 % (w/v) PEG 8000, 0.1 M CHES-NaOH pH 9.5, 0.2 M NaCl or 8 % (w/v) PEG 8000, 0.1 M Tris-HCl pH 8.5 (**Figure 7.3.3**).

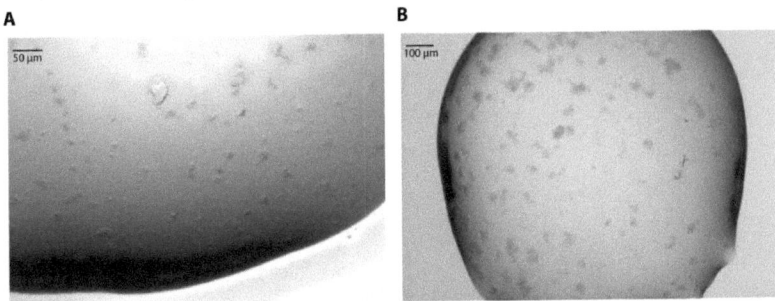

Figure 7.3.3 **Crystals of DDX1-674ΔHis-Tag with ATPγS.** DDX1-674ΔHis-Tag (without the N-terminal His-tag) at a concentration of 8 mg/ml = 105 µM, mixed with 8 x excess = 840 µM ATPγS, was used to set-up JCSG II screen (Qiagen). **A**, Crystal formation was observed after 2 days in condition A4 = 0.2 NaCl, 10% w/v PEG 8000, 0.1 M CHES pH 9.5. **B**, similar crystals formed in condition A12 = 8% w/v PEG 8000, 0.1 M Tris pH 8.5. Crystal plates were stored at RT.

Extensive screening around the two initial hit conditions was performed. Different PEG concentrations and different buffer pH values did neither improve crystal size nor lead to single crystals. Furthermore, the influence of additives on the crystals in the initial hit condition was tested, but did not show favorable effects. In the present form, the crystals are small and intergrown and can therefore not be used for X-ray diffraction experiments.

7.3.4 Co-crystallization of DDX1 in complex with RNA

A number of structures of DEAD-box proteins could be determined in complex with RNA[37] (PDB entries 2DB3, 2HYI, 2J0S, 3I5X). In these structures RNA stabilizes the helicase core by binding across the two RecA-like domains, bridging them and bringing them close together[23, Bono, 2006 #114, Sengoku, 2006 #78, Del Campo, 2009 #77], ensuring a rigid scaffold required for crystallization. The longest, stable DDX1 construct, DDX1-728 was complexed with 10-mer poly A RNA and used for crystal screening. For this experiment DDX1-728 protein (200 µM = 17 mg/ml) was mixed with 1000 µM 10-mer poly A RNA and incubated at 277 K for 20 min, before setting-up the crystallization trials using the JCSG core suite (Qiagen, Hilden, Germany).

Upon inspection of the plates immediately after setting the drops, heavy protein precipitation was observed in most wells of the JCSG screens.

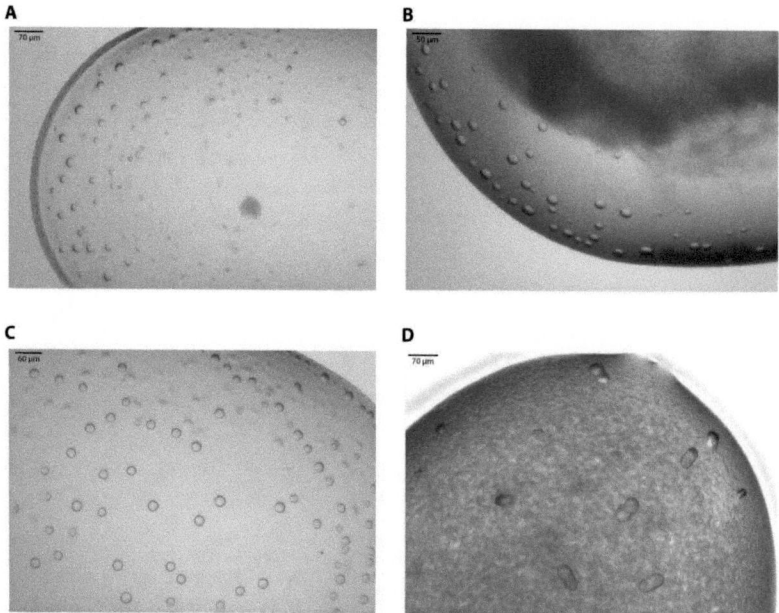

Figure 7.3.4 **Crystallization of DDX1 in complex with 10mer poly A RNA.** Exemplary conditions are shown that led to the growth of "bubbles" from protein precipitate after two days of incubation at room temperature. Conditions are from the JCSG core suite (Qiagen). **A**, B1 = 20 % w/v PEG 8000, 0.2 M MgCl$_2$, 0.1 M Tris pH 8.5. **B**, C5 = 1.4 M tri-Na citrate, 0.1 M HEPES pH 7.5. **C**, G7 = 0.2 M NaCl, 1 M tri-Na citrate, 0.1 M Tris pH 7.0. **D**, D5 = 0.2 M Mg formate.

Within two days of incubation at 297 K, "bubble"-like protein conglomerates grew from the precipitate in several conditions around pH 7.0 in the JCSG screens (**Figure 7.3.4**). In a condition with magnesium formate elongated shapes were found that showed some sharp edges (**Figure 7.3.4 d**). Screening around the hit-conditions did not help to transform the "bubbles" into protein crystals.

Furthermore, co-crystallization screening was also conducted with a different RNA substrate. The 14mer RNA that was used for the EMSAs (**see section 3.4.3**) (and was thereby confirmed to bind to DDX1) was employed to form a DDX1-RNA complex. To this purpose 200 µM (=17 mg/ml) DDX1-728 was mixed with 5 mM AppNHp, 10 mM $MgCl_2$ and 1200 µM 14mer RNA to form a ternary complex and used for screening in the JCSG core suite (Qiagen, Hilden, Germany). Two conditions of JCSG screen IV yielded small protein crystals (**see Figure 7.3.5**). In well A3, in 40 % (v/v) 2-Methyl-2,4-Pentanediol, 0.1 M CAPS pH 10.5 small rectangular crystals appeared after two days. In well B9, in 30 % (w/v) PEG 4000, 0.1 M Tris pH 8.5, 0.2 M $MgCl_2$ intertwined rods grew after one day. For both conditions fine-screens were set-up in 96-well format. For condition A3, 35 % (v/v) – 45 % (v/v) MPD and pH values from 9 to 11 were screened. For condition B9, 25 % (v/v) – 35 % (v/v) PEG 4000 and Tris pH 8.0 to 9.0 were screened. Crystals could not be reproduced in any of the fine-screens.

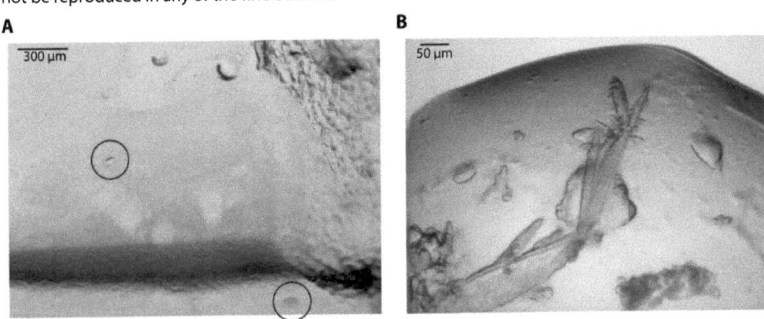

Figure 7.3.5 **Crystallization of DDX1 in complex with AppNHp and a 14mer RNA.** Two conditions of the JCSG core suite IV (Qiagen, Hilden, Germany) are shown that led to the growth of crystals. **A**, condition A3 = 40 % (v/v) 2-Methyl-2,4-Pentanediol, 0.1 M CAPS-NaOH pH 10.5 showed growth of small crystals, marked by blue circles. **B**, condition B9 = 30 % (w/v) PEG 4000, 0.1 M Tris-HCl pH 8.5, 0.2 M $MgCl_2$ led to the growth of intertwined rods.

In conclusion, co-crystallization of DDX1 with RNA substrate was extensively tested, but the amorphous aggregates or small crystals obtained, were unsuitable for X-ray diffraction experiments. Two different RNA sequences of varying length and protein in AppNHp- or apo- form were probed to stabilize a protein conformation that forms crystals. No growth of crystals, appropriate for structural studies was observed in any of the tested screens.

7.3.5 Transient crystal formation by using *in situ* proteolysis

To facilitate crystallization of stable fragments of DDX1, *in situ* proteolysis was employed. In this approach small quantities of a protease are added to the protein crystallization sample[282]. The protease slowly digests the protein and once stable fragments are produced, they can start to form crystals[282]. To this purpose a sample of recombinant DDX1 (construct DDX1-694) was spiked with protease and then immediately used to set-up crystallization screens, without assessing the efficacy of proteolysis, without purification of any degradation fragment and without stopping the proteolysis reaction[283]. Protein at 20 mg/ml (= 250 µM) was mixed with 0.2 mg/ml (= 8 µM) chymotrypsin (1:100 dilution) and used for screening at 297 K. Chymotrypsin preferably cleaves hydrophobic residues and cleavage sites are likely to be less frequent than sites for other proteases[282]. Crystal formation was observed after three days in JCSG screen I (Qiagen, Hilden, Germany), condition A1 (20 % (*w/v*) PEG 8000, 0.1 M CHES-NaOH pH 9.0) and A2 (20 % (*w/v*) PEG 6000, 0.1 M BICINE-NaOH pH 8.5) (**Figure 7.3.6**).

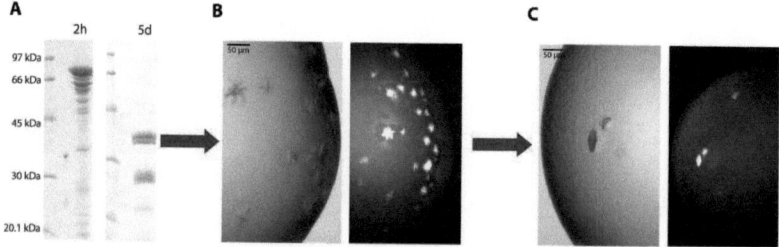

Figure 7.3.6 **Crystals obtained by in-situ proteolysis.** *In situ* proteolysis was used to obtain crystals with construct DDX1-694. **A**, SDS-gel showing the degradation of aliquots of the crystallization sample (DDX1-694 and chymotrypsin) after 2 hours and after 5 days. **B**, In an *in-situ* proteolysis experiment DDX1-694 at a concentration of 20 mg/ml was mixed with 0.2 mg/ml chymotrypsin and used to set-up JCSG I screen (Qiagen). Crystal formation was observed after 3 days in well A1 = 20% w/v PEG 8000, 0.1 M CHES-NaOH pH 9.0 (depicted here) and A2 = 20% w/v PEG 6000, 0.1 M BICINE-NaOH pH 8.5. Crystal plates were stored at RT. **C**, Optimized crystal found in a screen around the original hit conditions at 15% w/v PEG 6000, 0.1 M BICINE-NaOH pH 8.5. UV-pictures show Tryptophan-Fluorescence of protein-crystals.

Initial crystals consisted of multiple intergrown plates. By screening around the original hit condition, single crystals of a diameter of ~ 26 µm were obtained (**Figure 7.3.6 c**). Yet, crystal size was not sufficient for diffraction measurements and was resistant to optimization. After ten days, these crystals disappeared, which may be explained by the ever progressing proteolysis by chymotrypsin.

7.3.6 DDX1 without the SPRY insertion = DDX1ΔSPRY

Extensive partial proteolysis experiments were conducted with construct DDX1-694 and several commercially available proteases (**Figure 3.4, 7.3.7** and **7.3.8**). Peptide fragments, obtained in these partial proteolysis experiments were identified via MALDI-MS PMF (**Figure 3.4**). Based on these results, the entire SPRY domain was removed from DDX1 to facilitate structural studies on the helicase core.

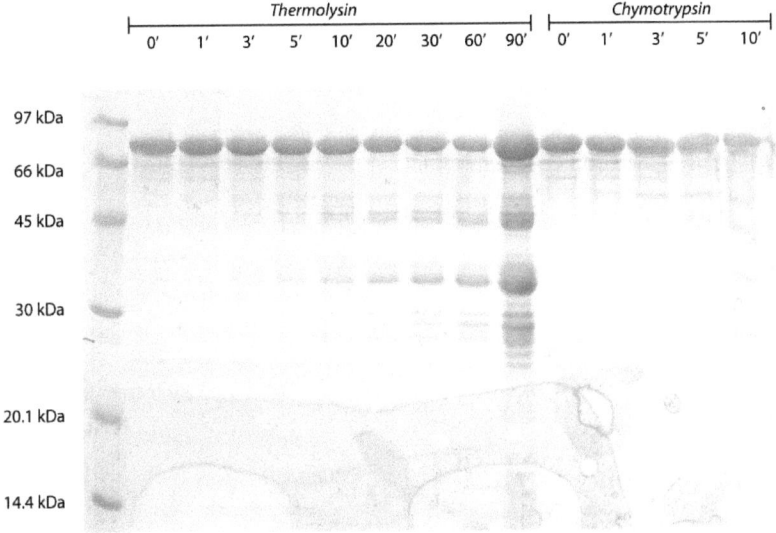

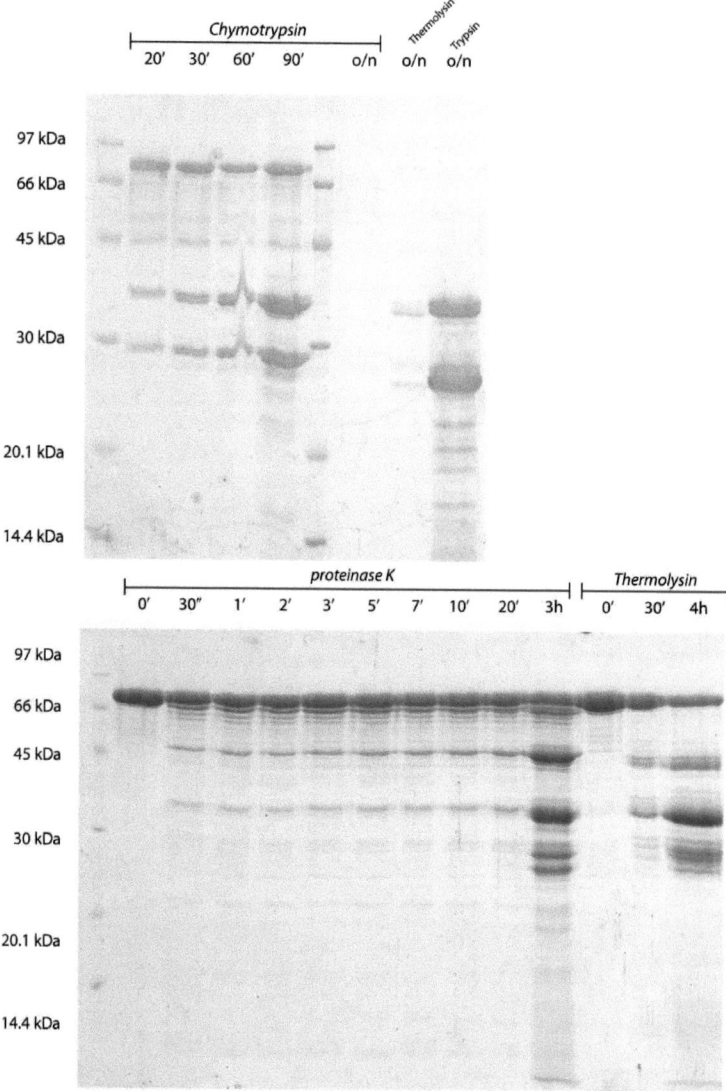

Figure 7.3.7 **Limited proteolysis of DDX1 with Thermolysin, Chymotrypsin and proteinase K.** Construct DDX1-694 was digested with Thermolysin or Chymotrypsin proteases (1 µg protease per 1600 µg protein) and reactions were quenched at the indicated time points (in min). Degradation bands were identified by MALDI-MS.

Appendix

Moreover, another partial proteolysis experiment was performed, where the products of a 24 h digest of DDX1-728 with trypsin were loaded on a Ni^{2+}-NTA affinity column (**Figure 7.3.8**). Fragments, corresponding to the SPRY domain (according to MALDI-MS PMF) were found in the flow-through, whereas RecA-like domains 1 and 2 were retained and could be eluted with an Imidazole wash. This further suggested that RecA-like domains 1 and 2 potentially interact tightly, but the SPRY domain does not and is separated from the core fold.

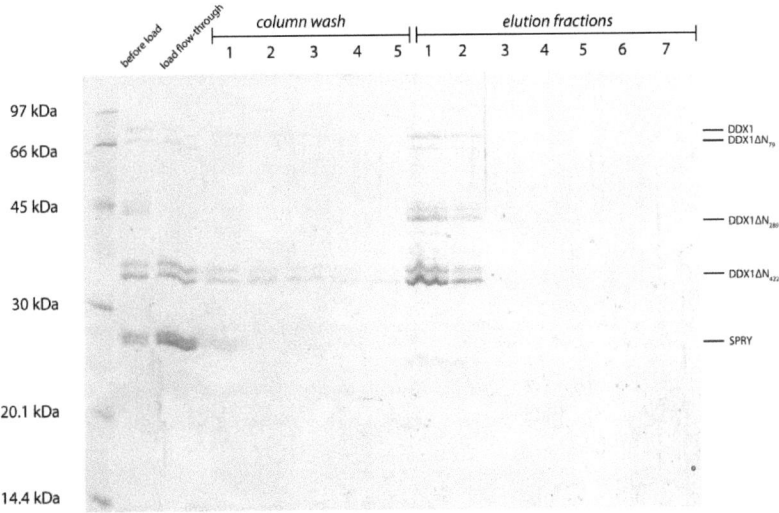

Figure 7.3.8 **Products of limited proteolysis, separated via Ni^{2+}-affinity chromatography.** Construct DDX1-694 was digested with trypsin over night. Digested protein sample was then loaded on a Ni^{2+}-affinity column. The column was extensively washed and then the bound protein was eluted by flushing the column with Imidazole. The lower two bands at around 28 kDa that are seen in the DDX1 digest lane, were not bound to the Ni^{2+}-NTAbeads. Those two bands correspond to the SPRY domain according to MALDI-MS. Apparently this domain does not extensively interact with the rest of the protein that stays bound to the Ni^{2+}-NTA, due to the presence of a His-Tag.

To create a DDX1 constructs that lacks the SPRY domain, initially the entire region coding for the SPRY domain was removed (Asp70 – Pro284). Domain boundaries were inferred from the MALDI-MS PMF results and refined based on sequence alignment of the entire sequence of human DDX1 with other helicases (**see Figure 3.2.12**). However, due to solubility problems of this initial construct, further constructs harboring less stringent SPRY domain excisions were designed.

Several of these constructs of human DDX1, lacking the SPRY domain, were screened for expression and solubility. Out of a few constructs that were expressed as stable proteins in *E. coli*, only one did not show aggregation on a size exclusion column. This construct, designated 'DDX1-694ΔSPRY' did comprise amino acids 1-68 and 248-694 of DDX1, i.e. the stretch comprising Asp70 to Phe247 had been removed. The construct was purified as WT DDX1 protein (**Figure 7.3.9**, see materials and methods, section 2.3.3).

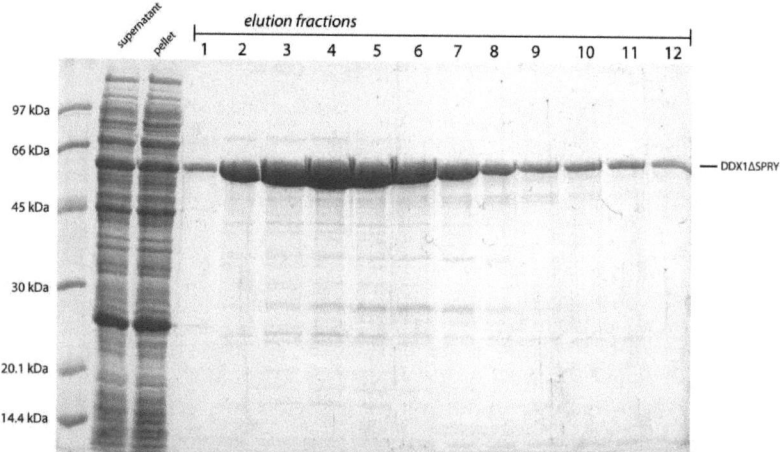

Figure 7.3.9 **His-Trap chromatography run with DDX1-694ΔSPRY.** The construct of DDX1 lacking the SPRY domain (DDX1-694ΔSPRY) was purified by Ni^{2+} affinity chromatography. Elution fractions were separated on a 15% SDS-gel.

The purified and concentrated construct DDX1-694ΔSPRY (see materials and methods, section 2.3.3) was tested for its folding and stability via CD measurements (**Figure 7.3.10**). The spectrum and the stability were similar to WT DDX1 protein (**compare Figure 3.3.1**).

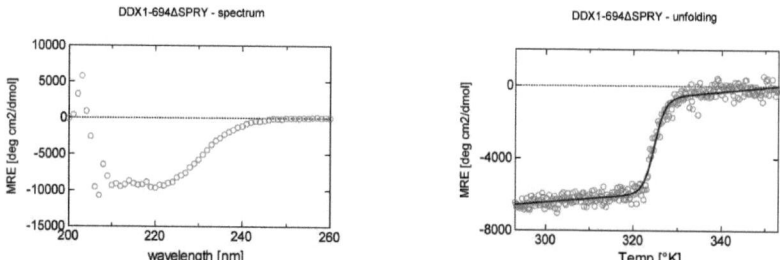

Figure 7.3.10 **Folding of DDX1-694ΔSPRY.** On the left side a CD spectrum of DDX1-694ΔSPRY is shown. CD signal measured with 5 µM protein was converted to mean-residue-ellipticity (MRE). On the right side, the temperature dependent unfolding curve as monitored by recording CD signal at 222 nm is shown. The data were fitted to a two-state-unfolding process. The fit yields a melting temperature T_m of 324.7 ± 0.10 K for DDX1-694ΔSPRY.

In contrast to WT DDX1 protein, purified recombinant construct DDX1-694ΔSPRY still displayed large heterogeneity after four column purification steps, detected in DLS measurements (**Figure 7.3.10**).

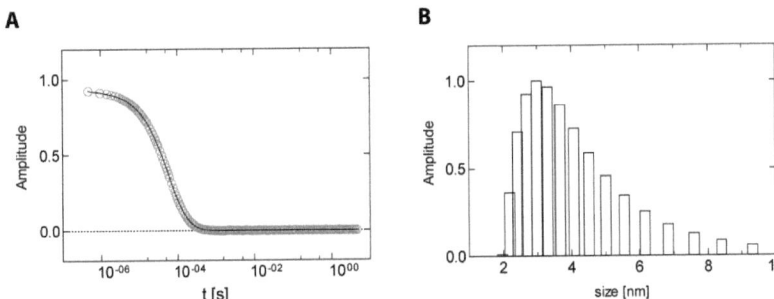

Figure 7.3.11 **DLS measurements with DDX1-694ΔSPRY.** The construct of DDX1 lacking the SPRY domain (DDX1-694ΔSPRY) was purified by 4 chromatography steps and then measured in a dynamic light scattering setup. **A**, combined autocorrelation function from 20 four second measurements is plotted in grey and a fit to the data in black. **B**, the distribution of the hydrodynamic radii by relative mass (amplitude of each bar indicates the percentage of the total mass of the sample) as obtained from the fit is shown. The fit did yield a peak in the mass-distribution of hydrodynamic radii at 3.99 nm (corresponding to a molecular weight of 87.13 kDa. Peak width is 50.4 % RSD, which corresponds to a heterogeneous sample.

After extensive optimization construct 'DDX1-694ΔSPRY' was obtained in quantities suitable for crystallization experiments. This construct did not show crystal formation in 'apo' and revealed large sample heterogeneity in DLS measurements (**Figure 7.3.11**). For this reason no crystal screening with RNA was conducted with the 'DDX1-694ΔSPRY' construct. Furthermore, construct "DDX1-694ΔSPRY" was tested for its ability to hydrolyze ATP and was found to be ATPase deficient.

7.3.7 Purification of DED1

A Plasmid coding for DED1 from *S. cerevisiae* - pET22b(DED1) was kindly provided by Isabelle Iost (Laboratoire ARN: Régulations Naturelle et Artificielle, INSERM U869, Université Victor Segalen, 146 rue Léo Saignat, 33076 Bordeaux, France).

DED1 was expressed in *E. coli* and purified as described in Iost and colleagues, 1999[59] and Fairman and colleagues, 2004[81]. Relatively pure DED1 protein could already be obtained by a single HisTrap Ni^{2+} affinity column run (**Figure 7.3.12**).

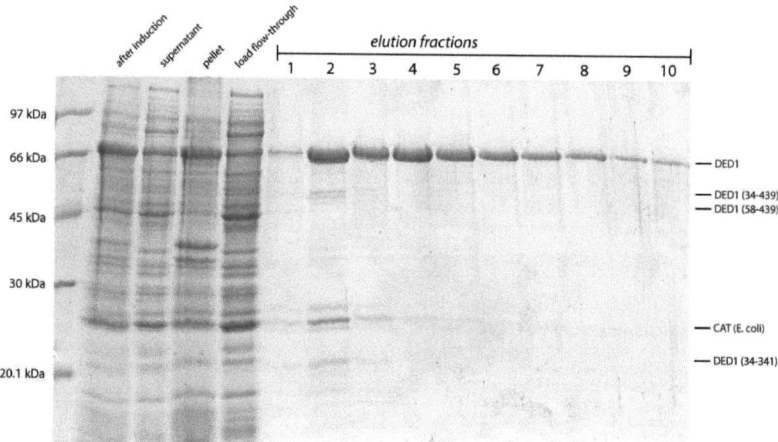

Figure 7.3.12 **Purification of the yeast protein Ded1p.** A construct of the yeast DEAD-box helicase Ded1p (in vector pET22b) that is involved in translation initiation was obtained from Isabelle Iost. The protein was expressed in E. coli and purified by initial Ni^{2+}-NTA affinity chromatography. Elution fractions from the Ni-beads were loaded on a 15% SDS-gel. Bands that were identified via MALDI-MS are indicated on the right.

7.4 Comments on DDX1 crystallization experiments

7.4.1 Crystallization of full-length DDX1

No crystals of the DDX1 helicase in its entity were obtained. As could be seen from limited proteolysis experiments (**Figure 7.3.7**), DDX1 consists of three major domains – RecA-like domains 1 and 2 and the SPRY domain – that are connected by flexible linkers. Individual domains were identified as relatively protease resistant elements, but the connecting linker regions get readily cleaved (**Figure 3.4**). The flexibility in the domain linkage and in the orientation of the domains relatively to each other probably precludes the formation of ordered protein crystals (**Figure 7.4.1** left side).

Appendix

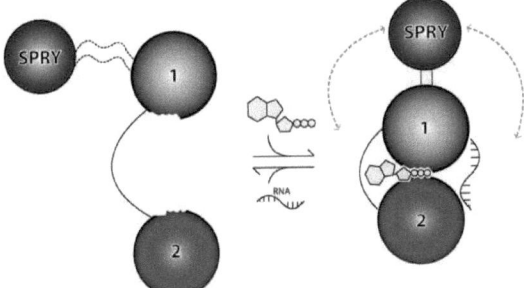

Figure 7.4.1 **Flexible linkage of the three DDX1 domains.** The three domains that form DDX1 are connected by flexible linkers. Upon binding of ATP and RNA, helicase domains 1 and 2 come close together and form the conformationally more rigid "closed" state. The orientation of the SPRY domain in this "closed" state is unknown and it might wobble between different conformations (indicated by dashed arrows). The SPRY domain is depicted in brown, RecA-like domain 1 in green and RecA-like domain 2 in blue.

For many other DEAD-box proteins well-diffracting crystals were obtained by complexing the proteins with RNA in co-crystallization experiments(Sengoku, Nureki et al. 2006). These nucleic-acid-bound structures show that the RNA binds on top of RecA-like domains 1 and 2 and keeps the helicase in a relatively rigid "closed" conformation(Mallam, Del Campo et al. 2012). In contrast to the proteins, whose structures could be solved by these means, DDX1 bears the large SPRY domain in its helicase core (**Figure 7.4.2** right side). This domain may also interact with RNA and probably influences DDX1 function. Since the domain insertion is large (~ 24 kDa) and positioned at a prominent location in the helicase core, it most probably has an influence on the conformation of the protein and on the relative orientation of the two helicase domains.

Two different RNAs were tested in crystallization experiments. No X-ray diffracting protein crystals of DDX1 in complex with RNA and ATP analogues were obtained. It could be shown by biophysical experiments though that the RNA oligomers used in crystal screening bind to DDX1 with relatively high affinity. Therefore, the inability to form crystals was attributed to domain flexibility or other structural determinants within the protein. Since the SPRY domain differentiates DDX1 from other DEAD-box helicases (that formed crystals with RNA) it is well possible that this domain insertion is one major element that hinders crystal formation. Therefore, a variant of DDX1 was engineered that lacks the SPRY domain. This construct did not crystallize, which might this time be due to folding problems of the mutilated protein variant. To speculate on the complete DDX1 protein and to build a homology model, it was presumed that the structure of DDX1 without the SPRY domain resembles the canonical DEAD-box helicase core. This assumption is valid due to significant sequence homology to other DEAD-box proteins (**Figure 3.2.11**). Several structures of such a helicase core have been deposited in the PDB[33, Hogbom, 2007 #12], which eased the search for a suitable modeling-template (**Figure 7.1.1**). Nevertheless due to the metabolic impact of human DDX1 a validation of the core structure by X-ray crystallography would be of interest.

7.4.2 Crystallization of DDX1 helicase domain 1

Extensive crystallization experiments have been conducted with construct DDX1-432, representing helicase domain 1/RecA-like domain 1. A structure of the separated RecA-like domain 1 of DDX1 would reveal the arrangement of the SPRY domain within the conserved helicase core. This would give important information on the possible influence of this domain on ATP-binding and ATP-hydrolysis, depending on its orientation to the ATPase active site. The SPRY domain might sterically obstruct the ATP-binding-site or it might interact with residues essential for ATP hydrolysis. Depending on the position of the SPRY domain with its surface charges, a structure would also give important clues on the role in binding RNA substrate. The SPRY domain may enlarge the RNA-binding site of helicase domain 1, formed by motifs Ia, Ib and Ic. A similar function has been observed for the C-terminal extension of yeast Mss116p[35].

The separated helicase domain 1, construct DDX1-432 did not crystallize. For many other DEAD-box proteins structures of the separated helicase domains have been solved[29, 40, 45]. For DDX1 again the SPRY insertion probably constitutes a major flexible element in helicase domain 1 that hinders crystallization. For further structural studies on DDX1 it might prove very helpful to crosslink the SPRY domain to the rest of the helicase. A crosslink could reduce flexibility and confer the ability to form ordered protein crystals. Such a crosslink has recently successfully been used for the crystallization of the chaperone DnaK in the ATP state[284].

Moreover components of the pentameric HSPC117 may interact with DDX1 directly via the SPRY domain (see section on SPRY structure). Addition of a DDX1-SPRY interacting component to the crystallization sample could also facilitate the formation of an ordered protein crystal by a great deal.

7.4.3 DDX1 conformational changes by nt binding – limited proteolysiss

Peptide fragments obtained by limited proteolysis of DDX1 in the presence or absence of nucleotides are identical (**Figure 3.3.2**). However, concluding from nucleotide binding and ATPase data, a conformational change upon binding of ATP is expected. There could be several reasons, why this conformational change is not reflected in the experimental limited proteolysis degradation pattern. First protease sites may not be influenced by the altered protein conformation and remain accessible for cleavage. Especially linker regions between the individual domains most probably display only minor positional changes between the "open"- and "closed"-state of the helicase. Second, AppNHp might not constitute an ideal replacement for ATP. Third, nucleotide might be constantly exchanged at the active site, which would make loops temporarily available for protease cleavage even in the presence of ATP analogues.

7.4.4 Structural studies on the HSPC117 tRNA ligation complex

Besides gaining structural insights into DDX1, major interest also lies in determining the three-dimensional structure of the entire human HSPC117 complex[119], which is involved in tRNA processing[114] (**see Introduction section 1.3.1**). Individual components of the HSPC117 complex were cloned and recombinantely expressed in *E. coli*. Proteins were either insoluble or showed aggregation at concentrations necessary for structural studies. Due to this lack in availability of recombinant HSPC117 complex components, crystallization experiments with the entire complex were not conducted. A possibility to rescue aggregated complex components might be co-expression. In this thesis mixing of bacteria lysates, expressing individual complex components has been tested. This combination of lysates did not assist the solubilization of aggregated protein (**Figure 3.1.5**). Actual bacterial co-expression of the proteins from the same vector may help to obtain soluble material from Fam98b and HSPC117. To explore which protein components can be expressed together, extensive interaction studies with the individual components are necessary (**Figure 3.1.6**). These interaction studies can already be conducted with the recombinant material from DDX1, CGI-99 and Fam98b. The establishment of expression and purification protocols for these proteins is described in this thesis (**see section 3.1.4**).

Moreover, the concomitant expression of all five complex components in bacteria might constitute another salvage strategy. Yet a further possibility to obtain material for structural studies on the complex could be expression in eukaryotic cells. The vector PFBDM, containing four of the complex components, DDX1, HSPC117, Fam98 and CGI-99 is available (cloned by Maike Gebhardt)[285]. This vector is suitable for the expression of eukaryotic protein complexes in insect cells[285]. Expression in insect cells offers additional advantages to bacterial expression, including post-transcriptional-modification of the peptides[286]. This constitutes the most promising approach to obtain material for determining the molecular structure of the HSPC117 complex in ensuing studies.

Since in the complete HSPC117 complex, all five protein components interact in a stoichiometric fashion and constitute a relatively stable particle[108], it is imaginable that they form X-ray diffracting crystals.

Of note the structure of the *Pyrococcus horikoshii* HSPC117/RtcB homologue 'PH1602-extein protein' has been solved and deposited in the PDB[287] (1UC2). This structure can guide the design of constructs for the recombinant expression of human HSPC117 in insect cells.

7.4.5 The SPRY domain and the HSPC117 complex-components may control the enzymatic cycle of DDX1

The SPRY domain provides a putative protein-protein interaction platform within DDX1 and primary candidates for docking are the factors of HSPC117 complex. Binding of co-factors to the SPRY domain can potentially modulate DDX1's enzymatic activity by influencing catalytically important residues via allosteric pathways. In addition, the SPRY domain seems to be flexibly connected to the helicase core and could change its relative orientation and thereby sterically block the active site. Furthermore, the putative function of the SPRY domain in extending the RNA binding surface of the helicase provides means of functional regulation of DDX1 (**Figure 7.4.2**).

The HSPC117 complex contains four proteins, additionally to DDX1, out of which three are poorly characterized. Complex components may act as nucleotide exchange factor or activate the ATPase cycle by stimulating RNA release. In addition, in the cell, they could direct DDX1 to a specific substrate (**Figure 7.4.2**).

Appendix

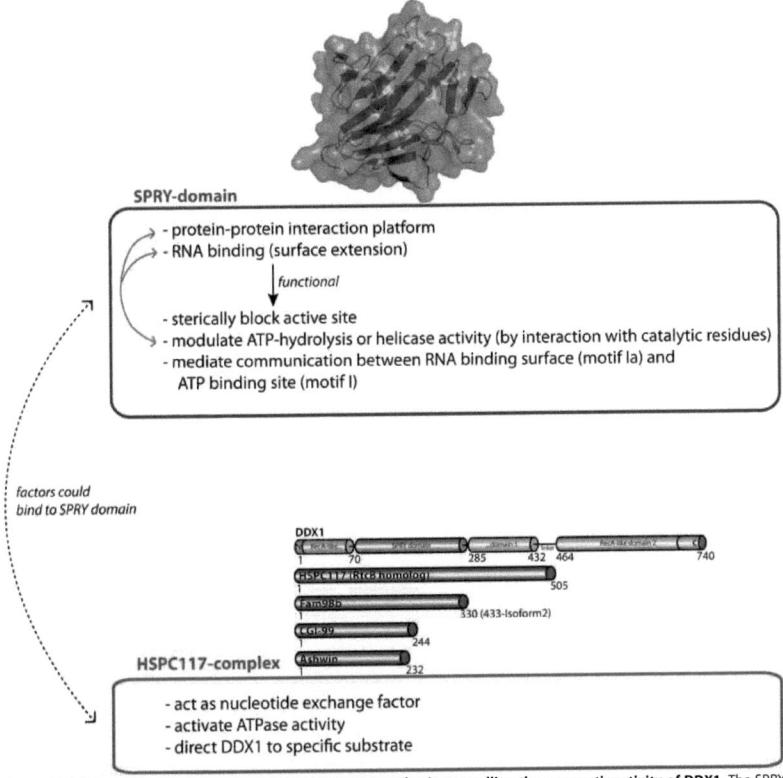

Figure 7.4.2 **Roles of the SPRY domain and the HSPC117 complex in controlling the enzymatic activity of DDX1.** The SPRY domain insertion within DDX1 is a putative protein-protein interaction platform, but may also extend the RNA interaction surface. It seems to be flexibly connected to the helicase core and in this way may conditionally block the active site. Due to the spatial proximity to the conserved helicase motifs, it could also modulate ATPase and helicase activity. The SPRY domain potentially interacts with factors of the HSPC117 tRNA ligation complex. HSPC117 complex component may act similar to factors in Dbp5- and eIF4A-DEAD box helicase complexes. They could constitute a nucleotide-exchange-factor and ATPase activating factor. They could also direct DDX1 to specific (structured) substrates.

7.4.6 Approaches for future studiess

The results from this work are the basis for deeper understanding of DDX1's metabolic role and the function of DEAD-box proteins in general. They enhance the working model for eukaryotic DEAD-box proteins by introducing synergistic effects of ATP and RNA binding and by potentially providing each molecule on its own with the ability to induce conformational changes of the protein. The data also have to be incorporated in the current mechanistic models of DDX1 to explain its cellular role and to devise *in vivo* experiments. Using the functional details reported in this thesis, upcoming studies can concentrate on several cellular aspects of DDX1.

The role of DDX1 in viral replication, specifically its interaction with HIV-1 REV protein[135], is definitely worthwhile further investigation. Affinity and binding constants for ATP, ADP and RNA can serve as the basis for an in-depth analysis for the hijacking function of viral transcripts and their modulatory propensity.

It could be rewarding to experimentally evaluate the role of extended surface A of the SPRY domain in RNA binding. Spectroscopic assays for assessing the RNA affinity have been established (**see chapter 3.3.7**), which paves the way for mutagenesis studies. Altering the charge of the identified positively charged patch by replacing residues could potentially change RNA affinity[288]. This, of course, would argue for a role of the SPRY domain in binding and tethering of RNA substrates to the helicase. Such a function could be conditional as it is dependent on the individual orientation of the SPRY domain relatively to the helicase core (**suppl. Figure 7.4.1**).

Research on the function of DDX1 in tRNA ligation can be further focused on structural studies of the HSPC117 complex. Here the structure of the SPRY domain of DDX1 is reported that provides a potential docking platform for the other HSPC117 complex protein components. Crystallization experiments could use cross-linking of the SPRY domain to either the DDX1 helicase core or to other protein factors in the complex. Arresting of the flexible SPRY domain could greatly facilitate protein crystal formation. The recombinant expression of some of the HSPC117 complex components, besides DDX1, is described in this thesis and with some additional optimization these proteins could be used for co-crystallization experiments. Remaining factors of the complex can be obtained e.g. from insect-cell expression. Prior to crystallization though, extensive interaction studies with the five protein components are necessary.

The general role of DDX1 in RNA processing is given a biochemical groundwork by the data in this thesis. The spectroscopic assays that were established for measurement of nucleotide affinity of DDX1 can be used for extensive mutagenesis studies. The implication of mutations in the conserved surface patch of the SPRY domain on RNA binding parameters can easily be accessed. Furthermore, the implications of those mutations on DDX1's enzymatic parameters like e.g. ATP hydrolysis are worthwhile to be investigated. A detailed mutational analysis of DDX1 should also include the conserved helicase motifs. This can provide an answer to the question, which residues are involved in establishing the cooperativity between ATP and RNA binding. Additionally, it will be very interesting to see, whether RNA binding may rescue ATP binding deficiencies of different Walker A and B motif variants.

Appendix

7.5 General figures on helicase structure and mechanisms

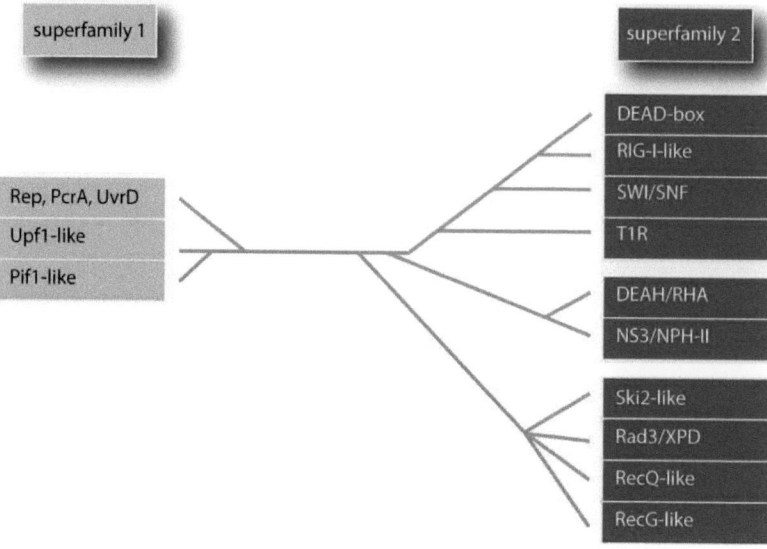

Figure 7.5.1 **Helicase superfamilies 1 and 2, classification.** The phylogenetic relationship between the SF1 and SF2 helicase families is shown, based on data from Jankowsky and Fairman-Williams 2010[12].

Appendix

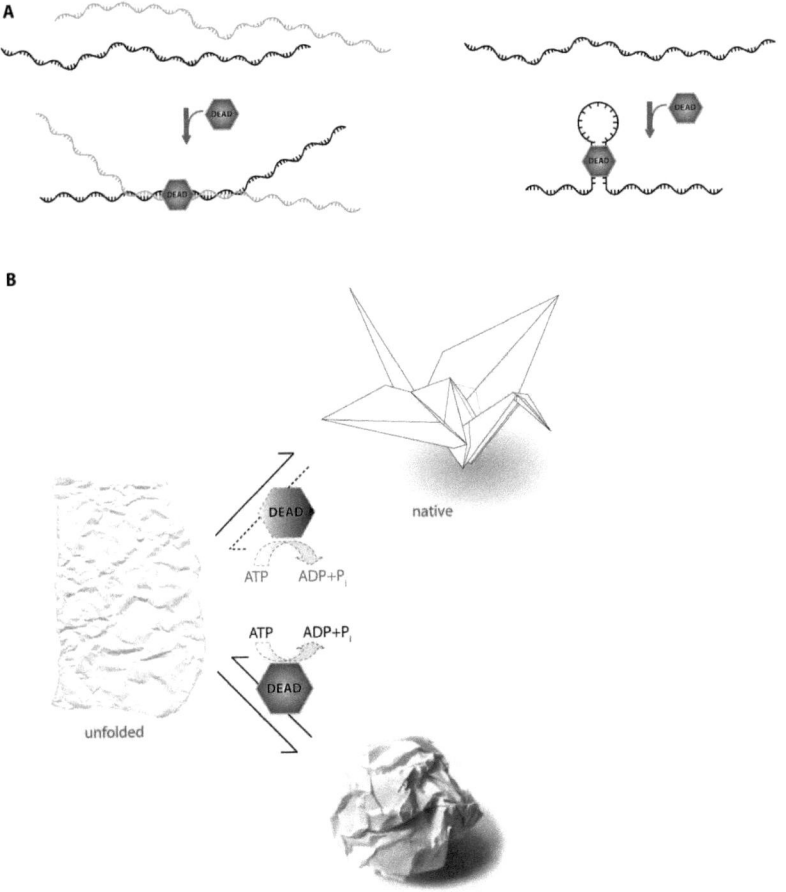

Figure 7.5.2 **Strand-annealing and "chaperone-like" functions of DEAD-box proteins. A**, DEAD-box proteins can accelerate the RNA duplex formation of two strands in solution without requirement of ATP. If strand annealing occurs intramolecular this can facilitate RNA folding. **B**, DEAD-box proteins can act as RNA-chaperones by non-specific unfolding of local structure. The less stable misfolded conformation is more readily "attacked" and unwound by DEAD-box helicases than the stable native structure. The helicase thereby shifts the equilibrium towards the native RNA state.

7.6 Details on the functions of DDX1

Table 7.6.1 Functions of DDX1 in RNA processing
(involvement as shown in previous studies)

Interaction	Localization	Implication	Reference
	colocalization with cleavage bodies in discrete nuclear foci	pre-mRNA processing	[91]
Binding to RNA cleavage stimulation factor (CstF-64)		pre-mRNA processing	[91]
	Component of mRNA transporting granules together with HSPC117 and CGI-99	mRNA transport	[115]
Interaction with heterogeneous nuclear ribonucleoprotein K (hnRNP K) and poly(A) RNA		transcription regulation	[109]
	Found in cytoplasmic stress granules in an RNP complex with YN-1 and MBNL1 (muscleblind-like)	splicing	[116]
Competes with 14-3-3 proteins for binding to KSRP(Kunde, Musante et al. 2011) (K homology splicing regulatory protein)		AU-rich element mediated decay (AMD)	[117]
Together with DDX21, DHX36 and TRIF it forms a receptor that recognizes viral dsRNA and elicits type I interferon (IFN) and cytokine responses. In the complex only DDX1 binds directly to viral RNA		Innate immune system (sensor of viral RNAs)	[137]
Binds to RelA and thereby enhances nuclear factor kappaB-mediated transcription		transcription regulation	[110]
	Is recruited to sites of DNA damage containing RNA-DNA structures	DNA repair	[111]

Posttranslational modification may also play a role in the regulation of DDX1 activity as *in vivo* it is phosphorylated in its helicase core and in the C-terminal region (**Figure 7.6.1**)[111, 289].

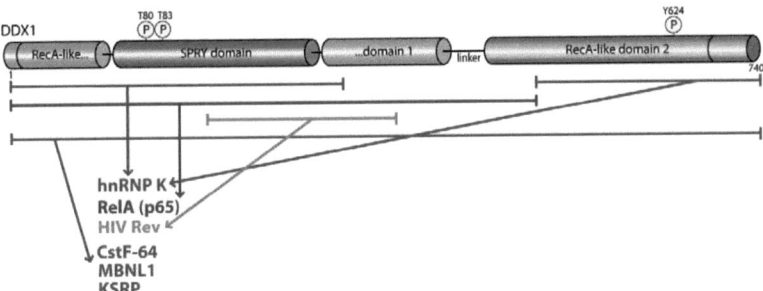

Figure 7.6.1 **Interaction of DDX1 with proteins involved in RNA procesing.** The regions of DDX1 that interact with other proteins are colored. Regions that interact with hnRNP K (heterogeneous nuclear ribonucleoprotein K) are coloured in purple, with RelA (p65) in red, with HIV Rev (regulator of virion) in green, and with CstF (Cleavage stimulation factor), MBNL1 (muscleblind-like), KSRP (K homology splicing regulator protein) in grey. Additionally phosphorylation sites are marked.

For the DDX1 homolog from Gallus gallus expression levels during development were investigated. Highest levels of DDX1 were found at early stages of development and tissue maturation was accompanied by a decrease in expression[290].

Table 7.6.2 **Role of DDX1 in formation and progression of tumors**
(DDX1 in cancer progression)

Type of Cancer	Involvement	Reference
Neuroblastoma	DDX1 is overexpressed; DDX1 is coamplified with *MYCN*	[79, 110, 111]
Retinoblastoma	DDX1 is overexpressed; DDX1 is coamplified with *MYCN*	[79, 110, 111]
testicular germ cell tumors	DDX1 activates transcription in testicular germ cell tumors and is a critical factor for testicular tumorigenesis	[112]
Wilms tumor	DDX1 is amplified in Wilms tumor	[291]
alveolar rhabdomyosarcoma cell lines	DDX1 is amplified	[114]
Breast cancer	Elevated DDX1 mRNA levels and elevated cytoplasmic DDX1 protein levels are associated with relapse and poor survival	[124]

Table 7.6.3 **Functions of DDX1 in viral replication pathways**
(hijacking of the body's own helicase)

Virus	Function	Reference
JC virus (polyoma virus)	high DDX1 expression renders cells susceptible to infection; DDX1 binds to the transcriptional control region of JCV together with CstF-64, which leads to productive expression and accumulation of viral transcripts	[98, 99]
Coronavirus	DDX1 binds to nonstructural protein 14 and colocalizes with viral RNA in the cytoplasm; Mutation of helicase motif II (DEAD) in DDX1 eliminated colocalization and replication of coronavirus, indicating an essential role for ATP hydrolysis and enzymatic activity of DDX1	[131]
Hepatitis C virus	DDX1 binds to the 3' untranslated-region (3'UTR) together with five other cellular proteins	[132]

| HIV type 1 | DDX1 is involved in HIV-1 mRNA trafficking; DDX1 interacts with the regulator of virion (Rev) protein and plays an important role in the nuclear export of unspliced mRNAs via the Rev-CRM1 pathway; DDX1 promotes oligomerization of Rev on Rev-response-element (RRE) mRNA, which is essential for the export; DDX1 can specifically bind to Rev in the absence of RNA and has an RRE RNA stimulated ATPase activity; RNA silencing experiments provide evidence that DDX1 is required for both Rev activity and production of HIV-1 virus particles; DDX1 cooperates with other DEAD-box helicases to modulate the HIV-1 Rev function | [102-107] |

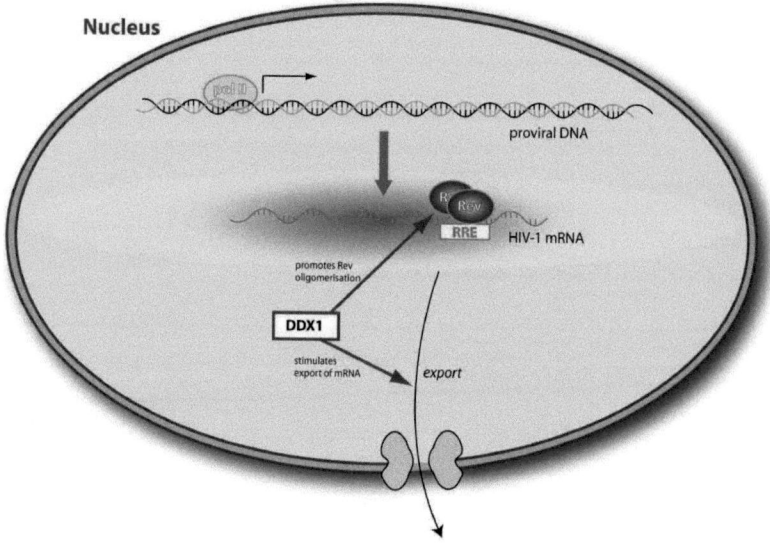

Figure 7.6.2 **HIV-1 utilizes DDX1 for export of viral mRNAs via the Rev pathway.** The Rev/RRE export pathway for unspliced viral mRNAs in HIV-1 is depicted. Rev protein binds to RRE on the mRNA assisted by DDX1. Oligomerization of Rev on RRE is essential for the export.

7.7 List of primers

Cloning primers

name	Sequence [5' -> 3']	application
DDX1_NheI_fwd	GGGAAGGGCTAGC<u>ATGGCGGCCTTCTCCG</u>	Amplify the DDX1 ORF, primer contains NheI site for cloning into pET28a
DDX1_NotI_rev	GGGATCGCGGCCGC<u>TCAGAAGGTTCTGAACAGCTG</u>	Amplify the DDX1 ORF, primer contains NotI site for cloning into pET28a
Fam98b_NheI_fwd1	TAGAT<u>GCTAGC</u>ATGAGAGGGCCGGAGCCGGG	Amplify Fam98b from PFBDM, primer contains NheI site for cloning into pET28a
Fam98b_XhoI_rev1	TAGAT<u>CTCGAG</u>TTACCAGGACATGAAAGAAAAT	Amplify Fam98b from PFBDM, primer contains XhoI site for cloning into pET28a
CGI-99_NheI_fwd1	TAGAT<u>GCTAGC</u>ATGTTCCGACGCAAGTTGACGGCTCTG	Amplify CGI-99 from PFBDM, primer contains NheI site for cloning into pET28a
CGI-99_XhoI_rev1	TAGAT<u>CTCGAG</u>TCATCTTCCAACTTTTCCCAGTCTGTG	Amplify CGI-99 from PFBDM, primer contains XhoI site for cloning into pET28a
ASW_NdeI_fwd2	TATAAT<u>CATATG</u>GCGGGGATGTGGGC GGTCGC	Amplify the ASW ORF, primer contains NdeI site for cloning into pET28a
ASW_BamHI_rev1	TAGAT<u>GGATCC</u>TTATCAGGGCCAAGTAACATGTTG	Amplify the ASW ORF, primer contains BamHI site for cloning into pET28a

Site-directed-mutagenesis primers (mutations are underlined, introduced nucleotides are italic)

name	Sequence [5' -> 3']	application
DDX1_genBamHI-pGEX_f	GCAGCCATATGG<u>A*TCC*G</u>CTAGCATGGCGGC	Generate a BamHI site at the N-terminus of DDX1 to clone it into pGEX-4T-1 vector
DDX1_genBamHI-pGEX_r	GCCGCCATGCTA<u>GC*GGA*T</u>CCATATGGCTGC	see above
DDX1_StopVal610_f	CGGTGTTCCTTATGTT<u>*GA*TAA*G*</u>TCACTCTGCC	Introduce double stop-codon after Val610 to generate pET28a(DDX1-610)
DDX1_StopVal610_r	GGCAGAGTGAC<u>*TT*ATC*A*</u>AACATAAGGAACACCG	see above
DDX1_StopArg648_f	CAACAGAAAAGAAAGGT<u>*GA*TAA*A*</u>CCATGTATGTAGC	Introduce double stop-codon after Arg648 to generate pET28a(DDX1-648)
DDX1_StopArg648_r	GCTACATACATGGT<u>*TTAT*C*A*</u>CCTTTCTTTTTCTGTTG	see above
DDX1_StopSer655_f	CCATGTATGTAGC<u>T*GA*TAA*G*</u>GAAAAGGG	Introduce double stop-codon after Ser655 to generate pET28a(DDX1-655)
DDX1_StopSer655_r	CCCTTTTCC<u>*TTAT*C*A*G</u>CTACATACATGG	see above
DDX1_StopGlu674_f	CATTTCTCAGGTTGAG<u>T*GA*TAA*A*</u>TAAAGGTACC	Introduce double stop-codon after Glu674 to generate

Name	Sequence	Purpose
DDX1_StopGlu674_r	GGTACCTTTAT*TTATC*ACTCAACCTGAGAAATG	pET28a(DDX1-674)
DDX1_StopArg694_f	CGGTCAGAAAAGG*IGATAA*CTGCTGGTGGTG	Introduce double stop-codon after Arg694 to generate pET28a(DDX1-694)
DDX1_StopArg694_r	CACCACCAGCAG*TTATC*ACCTTTTCTGACCG	see above
DDX1_StopHis728_f	GACATCTTTCCTGCAT*IGATAA*TACCTTCCTAAC	Introduce double stop-codon after His728 to generate pET28a(DDX1-728)
DDX1_StopHis728_r	GTTAGGAAGGTA*TTATC*AATGCAGGAAAGATGTC	see above
DDX1_StpLys432_f	CATGGGTTGACTTAAAA*IGATAA*GACTCTGTTCC	Construct of separated RecA-like domain 1 (including the SPRY domain)
DDX1_StpLys432_r	GGAACAGAGTCTTATC*ATTT*TAAGTCAACCCATG	see above
DX1SPRY_1gBamH-f	GTTTATGAAACTCTGAAAG*GATC*CCAGGAAGGCAAAAAAG	Gen BamHI site after Lys69 to remove SPRY domain
DX1SPRY_1gBamH-r	CTTTTTTGCCTTCCTGGG*ATC*CTTTCAGAGTTTCATAAAC	see above
DX1SPRY_2gBamH-f	CAAACAAAGTTTGGA*TCC*AATGCTCCGAAAGC	Gen BamHI site before Asn285 to remove SPRY domain
DX1SPRY_2gBamH-r	GCTTTCGGAGCATTGG*ATC*CAAACTTTGTTTG	see above
DX1SPRY_3gBamH-f	CGGTGAAGAGGGATCC*AAGTTTCCACCAAAAG	Gen BamHI site before Lys248 to remove SPRY domain
DX1SPRY_3gBamH-r	CTTTTGGTGGAAACTT*G*GATCCCTCTTCACCG	see above
DX1_WlkrA_KtoA_f	GAAACAGGAAGTGGC*G*CAACTGGTGCTTTAG	Mutate Lys52 to Ala (Walker A)
DX1_WlkrA_KtoA_r	CTAAAAGCACCAGTTG*C*GCCACTTCCTGTTTC	see above
DX1SPRY_4gXhoI-f	CAAACAAAGTTTCTG*AGA*ATGCTCCGAAAGCTC	Generate XhoI site downstream of the SPRY domain to clone it as BamHI-XhoI fragment
DX1SPRY_4gXhoI-r	GAGCTTTCGGAGCATT*C*AGGAGAAACTTTGTTTG	see above
SPRY_Gly84BamH-f	GAAAAACAACAATTAAA*GGATCC*GCTTCAGTGCTG	Generate BamHI site after Gly84 to truncate N-term of SPRY domain
SPRY_Gly84BamH-r	CAGCACTGAAGCGG*ATCC*TTTAATTGTTGTTTTTC	see above
SPRY_Ser100BmH-f	CATATGACAGAG*GATCC*GCTTTTGCAATTGGG	Generate BamHI site after Ser100 to truncate N-term of SPRY domain
SPRY_Ser100BmH-r	CCCAATTGCAAAAGC*GGATC*CTCTGTCATATG	see above
SPRY_StopAla261_f	CTCTTTCCAAGGCA*IGATAA*GGTTACATTG	Generate double stop-codon after Ala261 to truncate C-term of SPRY domain
SPRY_StopAla261_r	CAATGTAACC*TTATC*ATGCCTTGGAAAGAG	see above

Sequencing primers

Name	Sequence	Purpose
DDX1_NheI_fwd	GGGAAGGGCTAGC*ATGGCGGCCTTCTCCG*	Sequence N-terminus of DDX1
DDX1_seq_1	TCTTTCCAAGGCACCGGATG	Sequence middle region of DDX1
DDX1_NotI_rev	GGGATCGCGGCCGC*TCAGAAGGTTCTGAACAGCTG*	Sequence C-terminus of DDX1

Appendix

pET_bone_for	ATGCGTCCGGCGTAGAGGATC	Sequence pET28a MCS from N-terminus
pET-28b_back_r	GCGGTGCAAGCGGGCCGAAAGGTTCGAATATACG	Sequence pET28a MCS from C-terminus

7.8 Numerical data analysis with Dynafit

To extract RNA binding affinities of DDX1, all equilibrium titration data for the binding of mant-labeled and unlabelled ATP/ADP in the presence and absence of RNA was fitted globally by numeric iteration using Dynafit[198] (see results section 2.4.7, Biokin ltd). The Dynafit input script file (in *.txt format) contains the binding mechanism, that gets converted to a system of equations to describe the alterations of the different species. The parameters of the equations are optimized to obtain a minimal deviation of the simulations to the input data traces.

The script file first contains the tasks, meaning the form of data (equilibrium titration or temporal progress or …), the actual operation mode (here "fit") and the binding model.

Then the mechanism and the corresponding rate constants are defined.

The file contains estimated initial values for the rate constants to be optimized, indicated by the "?" behind the values. By putting "??" behind a value the programme does do a sensitivity analysis and calculates the confidence intervals for each parameter.

Amplitudes are represented by response parameters that correspond to the change in signal upon formation of 1 unit (µM in this case) of the defined signal-producing species (i.e. the difference between basal signal and the signal of a defined species/complex). Response parameters were fitted globally as parameter "binding1" and are therefore named in the script file directly next to each input dataset, thus the [response] line is empty, but the [parameters] line contains the global response parameter "binding1". Concentrations are named directly next to the input data so the [concentrations] line is also empty.

```
;----------------------------------------------
;mant-dADP complexed with DDX1 (=H), then a fraction is displaced by adding ATP
;further displacement is achieved by titrating 10mer polyA-RNA (see 19.06.13)
;
;----------------------------------------------

[task]

    data  = equilibria
    task  = fit
    ; confidence =    monte-carlo
    model = KEAGlobalEqu

[mechanism]
```

```
H + mant <===> Hmant      :   Kdmant   dissoc
H + ATP  <===> HATP       :   KdATP    dissoc
Hmant + RNA <===> HmantRNA :  KdRNA    dissoc
HATP + RNA <===> HATPRNA  :   KdRNA2   dissoc
H + RNA <===> HRNA        :   KdRNA    dissoc

HRNA + mant <===> HmantRNA :  Kdmant   dissoc
; HRNA + ATP <===> HATPRNA :  Kd2      dissoc

[constants]

   Kdmant  = 0.088
   KdATP   = 129 ??

   KdRNA   = 10 ??
   KdRNA2  = 0.01 ??




[responses]

[equilibria]

[concentrations]

[parameters]

   Binding1 = 2500 ?
   Binding2 = 2500 ?


[data]

   directory ./current/dynafit_input_data/10merRNA_titration

variable    RNA

; Kd2 = 1 * (KdATP * KdRNA2) /KdRNA


file 50_µM_ATP_10mer.txt
concentration mant = 0.2, H = 1 , ATP = 50
response Hmant = 1 * Binding1
response HmantRNA = 1 * Binding1
offset = 6000 ? (1000..9000)



file 100_µM_ATP_10mer.txt
concentration mant = 0.2, H = 1 , ATP = 100
response Hmant = 1 * Binding1
response HmantRNA = 1 * Binding1
offset = 6000 ? (1000..9000)


file 200_µM_ATP_10mer.txt
concentration mant = 0.2, H = 1 , ATP = 200
response Hmant = 1 * Binding1
response HmantRNA = 1 * Binding1
offset = 6000 ? (1000..9000)



file 400_µM_ATP_10mer.txt
concentration mant = 0.2, H = 1 , ATP = 400
```

Appendix

```
response Hmant = 1 * Binding1
response HmantRNA = 1 * Binding1
offset = 6000 ? (1000..9000)


;-----------------------------------------------------------
;this is where previous titration data is included in the global fit
;I'm not sure, whether it works this way
;
;-----------------------------------------------------------

directory ./current/dynafit_input_data/mant-dADP_binding

variable    H

file mant_dADP_binding.txt
concentration mant = 0.05
response Hmant = 1 * Binding1
offset = 6000 ? (1000..9000)



directory ./current/dynafit_input_data/competition_binding_ATP

variable    ATP

file ATP_competition_1.txt
concentration mant = 0.2, H = 1
response Hmant = 1 * Binding1
offset = 6000 ? (1000..9000)



directory ./current/dynafit_input_data/competition_binding_ATP_RNA

variable    ATP

file 8_µM_10mer_polyA.txt
concentration mant = 0.2, H = 1 , RNA = 8
response Hmant = 1 * Binding1
offset = 6000 ? (1000..9000)



;mesh    linear from 0 to 30 step 0.15

[output]

    directory ./current/output

[[Settings]]
| Interrupt = 500

Level         = 95.0
Minsteps              = 40
MaxSteps              = 100
OnlyConstants   = n
DetailedReport  = y
```

Figure 7.8.1 **Dynafit script file**
Dynafit script file as used for the numerical data fitting. Parameters fitted globally are named as constants, whereas the semi-global fitted parameters are named in the [parameters] section.

The [parameters] optimized during the fit contain the response parameters for each data-set. Next the input data is defined with the specific directory within the biokine folder. The initial species concentrations, the complex constituting the response parameter and the actual parameter for the response are defined.

The offsets were fitted locally as indicated by their positioning next to each data set (the "?" behind the values as the fitting command). The values in brackets give the constraints, i.e. the upper and lower limits for the parameter. Last the output directory for the result files is specified and no filter is applied to the data sets (since they have already been formatted and scaled manually before).

The result of the Dynafit run is a parameter file containing the optimized parameters with their respective standard errors and the simulated curves.

Note that all concentrations in the script file have to correspond to the same concentration, which is µM in this case, leading to an output in µM. The input data has to be in *.txt files, not containing any space in the file-names.

7.9 List of abbreviations

aa	amino acid
ADP	adenosine-5'-diphosphate
ATP	adenosine-5'-triphosphate
AppNHp	adenosine-5'-[(β,γ)-imido]triphosphate
ATPγS	adenosine-5'-(γ-thiol)-triphosphate
APS	ammonium persulfate
a.u.	arbitrary units
bp	base pair
BSA	bovine serum albumin
CD	circular dichroism
C-terminus	carboxy terminus
Δ	differential or delta, i.e. without
DDX1	DEAD-box helicase 1
dpi	dots per inch
DTE	dithioerythritol
DNA	desoxyribonucleic acid
mRNA	messenger-ribonucleic acid
E. coli	*Escherichia coli*
EDTA	ethylene diamine tetraacetic acid
g	gram
g	gravity
h	hour
HEPES	4-(2-hydroxyethyl)-1-piperazineethanesulfonic acid
IPTG	isopropyl-β-D-thiogalactosid
K	degree Kelvin
K_d	dissociation constant
K_m	michaelis-menten constant
k_{on}	on-rate constant
k_{off}	off-rate constant
kDa	kilo Dalton
l	liter
LB	Lysogeny broth (incorrect Luria Bertani)
LDH	lactate dehydrogenase

M	molar
mA	milliampere
mant-	N-methylanthraniloyl-
mant-dADP	2'-deoxy-3'-O-(N'-methylanthraniloyl)adenosine 5'-O-diphosphate
mant-dATP	2'-deoxy-3'-O-(N'-methylanthraniloyl)adenosine 5'-O-triphosphate
min	minute
MWCO	molecular weight cut off
mdeg	millidegree
NADH	Nicotinamide adenine dinucleotide, reduced form
n.a.	not available
N-terminus	amino terminus
n.d.	not determined
nm	nanometer
OD	optical density
o/n	over night
ORF	open reading frame
PCR	polymerase chain reaction
PDB	protein data bank (www.rcsb.org)
PEP	phosphoenol pyruvate
PK	pyruvate kinase
PMSF	phenylmethylsulphonyl fluoride pyrene N-(1-pyrene)-maleimide
rcf	relative centrifugal force
rpm	rounds per minute
RNA	ribonucleic acid
rRNA	ribosomal RNA
RT	room temperature
s	seconds
SDM	site-directed mutagenesis
SDS	sodium dodecyl sulfate
t	time
T	temperature
TEMED	N,N,N',N'-tetramethyethylenediamine
Tris	tris(hydroxymethyl)-amino-methane
tRNA	transfer RNA
U	unit
UV	ultraviolet
(v/v)	(volume/volume)
w/o	without
WT	wildtype
(w/v)	(weight/volume)
Yeast	saccharomyces cerevisiae

I want morebooks!

Buy your books fast and straightforward online - at one of the world's fastest growing online book stores! Environmentally sound due to Print-on-Demand technologies.

Buy your books online at
www.get-morebooks.com

Kaufen Sie Ihre Bücher schnell und unkompliziert online – auf einer der am schnellsten wachsenden Buchhandelsplattformen weltweit!
Dank Print-On-Demand umwelt- und ressourcenschonend produziert.

Bücher schneller online kaufen
www.morebooks.de

OmniScriptum Marketing DEU GmbH
Heinrich-Böcking-Str. 6-8
D - 66121 Saarbrücken
Telefax: +49 681 93 81 567-9

info@omniscriptum.com
www.omniscriptum.com

Printed by Books on Demand GmbH, Norderstedt / Germany